THE
TINSMITH'S HELPER
AND
PATTERN BOOK

WITH USEFUL RULES, DIAGRAMS AND TABLES

BY H.K. VOSBURGH

SIXTH EDITION

ASTRAGAL PRESS
Lakeville, Minnesota

Reprinted in cooperation with
the American Tinsmith Museum
of Sturbridge, Massachusetts,
from an original volume in the
library collection.

Copyright © 1994 by Astragal Press
All rights reserved

Library of Congress Catalog Card Number: 94-78584
ISBN 10: 1-879335-56-5
ISBN 13: 978-1-879335-56-1

Published by
Astragal Press
An Imprint of Finney Company
8075 215th Street West
Lakeville, Minnesota 55044

www.finneyco.com
www.astragalpress.com

Printed in the United States of America

NOTE TO THE READER.

The materials and methods described in this book are from an earlier era when safety was of less concern. Some of these may now be considered dangerous, so care and good judgment should be exercised. The Astragal Press has neither tested nor endorses them and offers this reprint solely as a source of historical information.

INTRODUCTORY.

The first edition of this book appeared in 1879, and since then it has had a continued and increasing sale. The author, H. K. Vosburgh, knew from experience the needs of the practical tinner and prepared a book in which a number of simple patterns were described in the plainest way. In preparing the fourth edition the greater part of the work of Mr. Vosburgh has been scrupulouly preserved, but quite a number of important changes have been made by William Neubecker, pattern cutting expert, mainly in the direction of greater accuracy and in accordance with modern methods.

INDEX.

DIAGRAMS AND PATTERNS.

To Find the Center of an Arc.

Fig. 1.

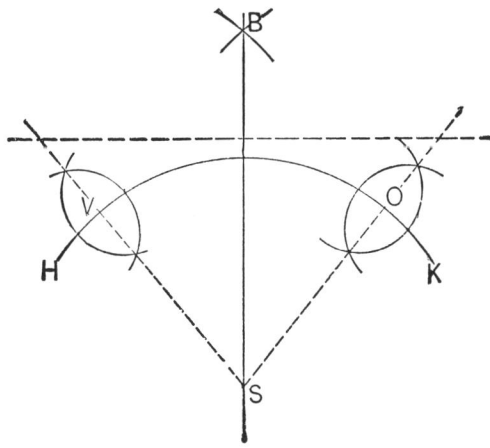

Let H K represent the given arc. Span dividers any convenient radius and describe small arcs, as V O. Draw lines through them, as shown by dotted lines, and the intersection, S, will be center sought.

To Describe an Octagon Within a Given Square.

Fig. 2.

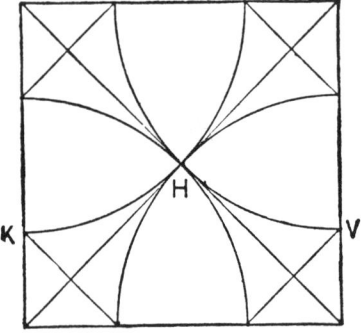

Draw diagonal lines from corner to corner and the intersection is the center H. With the compasses set to a radius from center to corner, and one foot set successively at each corner, describe the arcs, as shown. The points at which they cut the square, as K V, will be the corners of the octagon. Draw lines from point to point to complete the figure.

To Describe an Octagon Within a Given Circle.

Fig. 3.

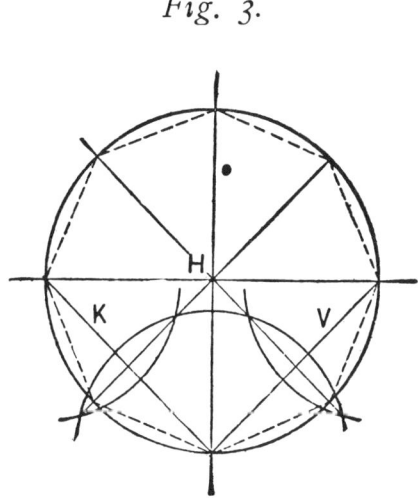

Draw lines at right angles passing through the center H. This divides the circle into four equal parts, which need only to be subdivided into equal parts again to form the corners for the octagon. This may be easily done by drawing the lines K V, bisecting, as shown, and drawing lines to the circle.

The bottom will correspond in size to the size of the circle or square. Remember to allow for burr and double seam.

To Describe Breasts for Cans.

Fig. 4.

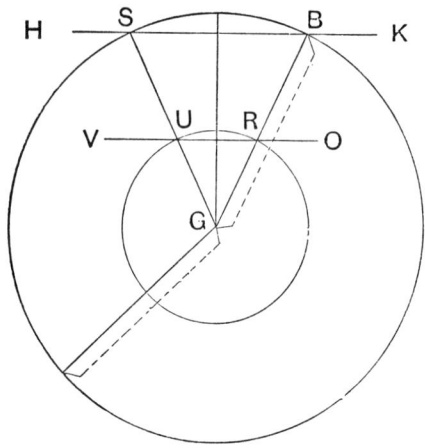

Draw horizontal line H K, another parallel to it, V O, making the distance between the desired hight of breast. On H K lay off diameter of can, as S B. On V O, size of opening as U R, produce lines B R, S U, until they cross G. Span dividers from G to S, describe outer circle. G to U, describe inner circle. Set off outer circle equal to the diameter of the can B S. Starting at B, draw line from G, allowing for locks, as shown by dotted lines. *Reference can be made to the circumference table.*

Can Breasts.

Fig. 5.

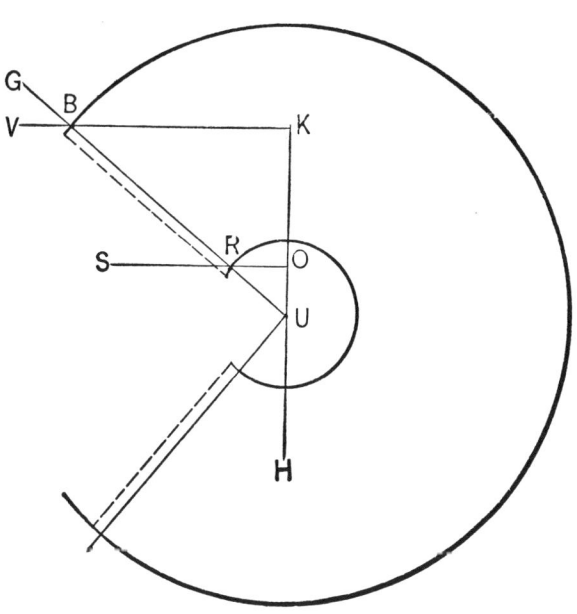

Draw the two horizontal lines, K V and O S, and perpendicular to them the line K H. Set off on line K V from the point K one-half the diameter of the can. On O S the point R is one-half the diameter of the opening. Produce the line U G, touching the points B and R, until it intersects H K. From U as center, with the radius U B, describe the outer circle. With the radius U R, the inner. Then span from K to B and step six times on large circle to obtain size of breast. Draw line to center and allow for locks, as shown by dotted lines.

Can Breasts.

Fig. 6.

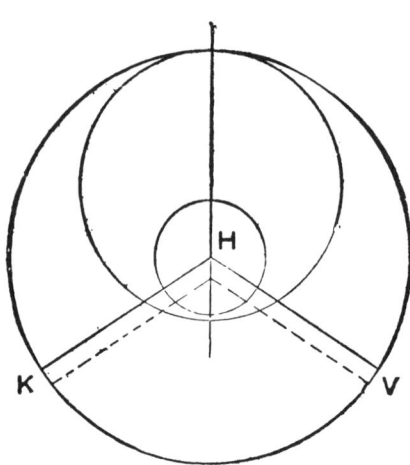

Describe circle size of can. Draw line through center H. Span dividers three-fourths of diameter and strike circle K V. Span to diameter of can and step three times on large circle.

Draw line from center to points K V, allowing for edges and locks. For more or less pitch make circle K V larger or smaller.

Small circle in center for opening in top. Hoods and pitched covers may be cut by same rule.

Pattern for Cone.

Fig. 7.

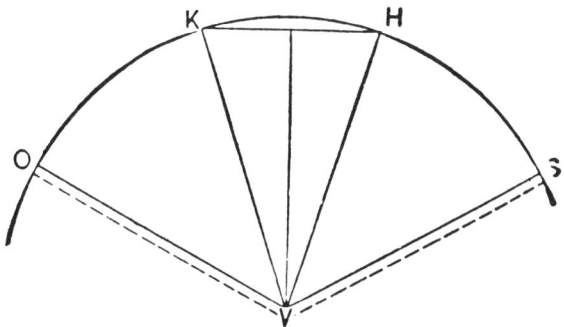

H K V represents a cone for which an envelope is wanted.

Span the dividers from V to H and describe the arc O S. Set off the arc equal in length to the circumference of the required cone. Draw the lines V O and V S, allowing for locks or laps, as shown by the dotted lines.

For the circumference, refer to the tables or obtain by some of the rules. By using the rules familiarity with them is obtained, which is desirable.

To Describe Pattern for Flaring Vessels.

Fig. 8.

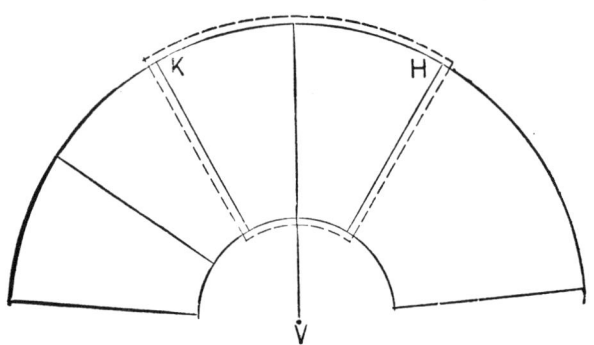

For example, it is desired to describe pattern for pail 12 inches in diameter at top, 9 inches at bottom and 9 deep.

Take the difference between large and small diameters (3 inches) for the first term, the hight for the second and the large diameter for the third, thus, 3 : 9 : : 12.

12 x 9 ÷ 3, this gives radius by which the pattern may be described. Span the dividers (or use beam compasses, piece of wire, straight edge or any convenient device) 36 inches and strike large circle. With radius less the slant hight of pail strike small circle. Ascertain the cir-

cumference required and divide by the number of pieces to be used. Lay off on outer circle and draw lines to center, as H K V.

Allow for locks, burr and wire.

To Cut Hood for Stove Pipes.

Span dividers size of pipe, describe circle, cut in to center, lap over and rivet.

To Describe Patterns for Flaring Tinware.

Fig. 9.

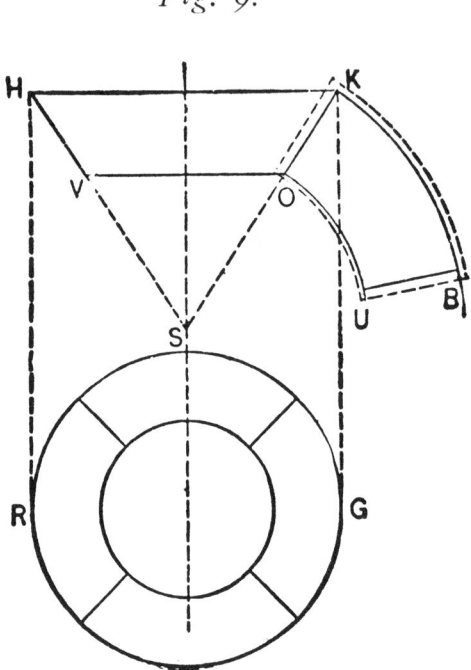

By this figure and rule can be drawn any article of flar-
ing tinware of any diameter, large or small. It is a rule
of more extensive application than any other for getting
correct patterns for frustums of a cone. It is the foun-
dation for all curved work, cornice, bevels, chamfers, etc.

H K V O represents the elevation of an ordinary tin
pan, constructed in four pieces, 15½ inches in diameter at
the top. Below the elevation is shown the same in plan;
the pan is a frustum of a cone, and if the sides of the pan

were continued down until they intersected at S, as shown, the cone would be complete. The radius of the envelope of the cone must be either S H or S K. To describe the section of the frustum which is required, place one foot of the dividers at the center S, and with the radius S H describe the arc K B. With the radius S V describe O U. This gives the width of pattern and the proper sweep.

To get the length of the piece, refer to the table of circumferences or find, by the rules given, the circumference of the article, which in this case is 48⅝ inches. There being four pieces, divide by four, which gives 12 5-32 inches; span the dividers 1 inch, step off the 12 and add the fraction.

Draw line from center S to point last ascertained. For locks, wire edge and burr allowance must be made.

The Old German Rule for Patterns for the Cone.

Fig. 10.

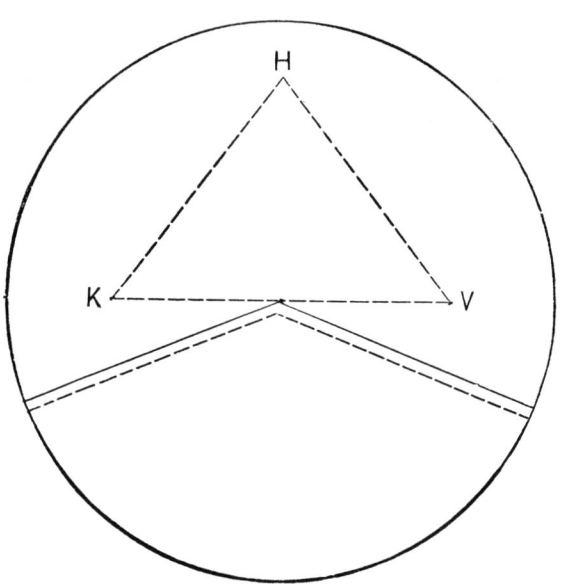

Take the slant hight of the cone H K as a radius, and describe a circle. Divide the diameter of the base of the cone K V into seven equal parts and set off a space equal to twenty-two of these parts on the circle already struck. From the extremities thus measured off draw lines to the center.

Allow for locks.

Frustum of a Cone.

Fig. 11.

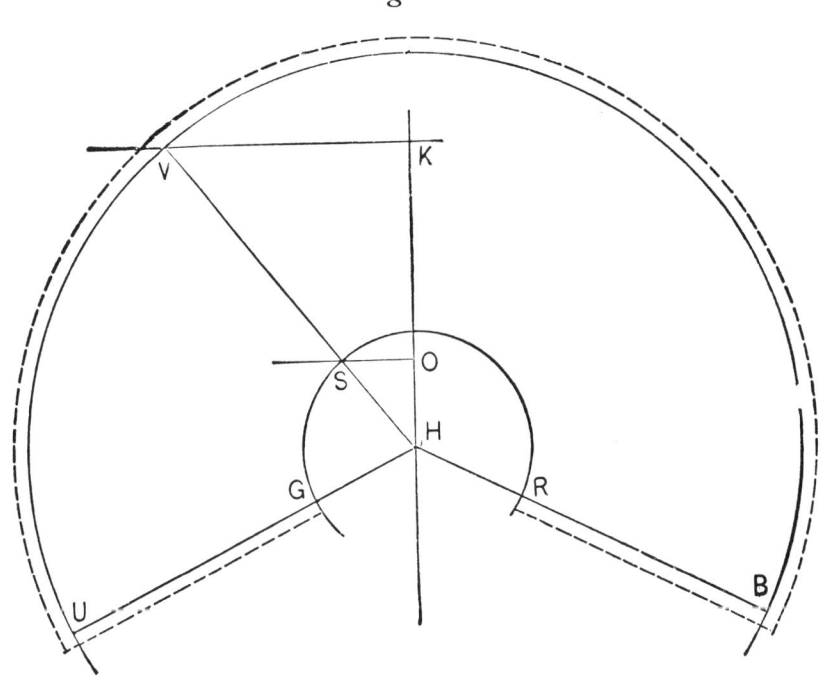

Lay the square on your sheet and construct the right angle H K V. Draw line O S parallel to K V, making the distance K O the altitude. On these lines lay off one-half the diameter of the large and small ends. Draw line through points V and S until they intersect at H; then, with H as the center, describe the semicircles B U, R G. Lay off circumference of large end on line B U and draw lines to center H. Must allow for all edges. For two sections take one-half of the piece, allowing edges on piece used for pattern.

Flaring Vessel in Three Pieces.

Fig. 12.

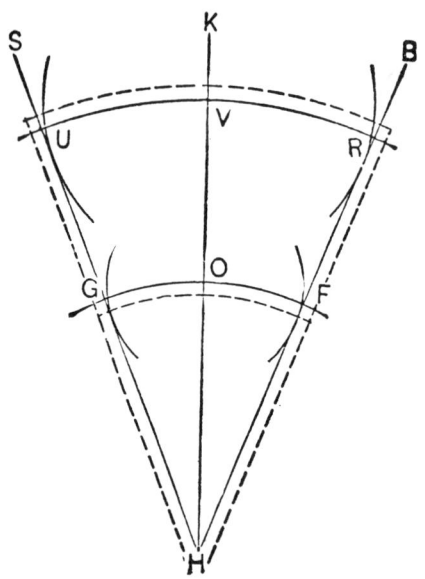

Draw line H K; perpendicular to it, lines parallel to each other apart the hight of vessel. With the intersections, as V, O for centers, describe circles size of top and bottom of vessel. Draw lines S H and B H touching on circles, and at intersection H as center, with the radius H V, describe the segment U R; with the radius H O, the segment G F. Allow for locks, as shown by dotted lines.

Frustum of a Cone

Fig. 13.

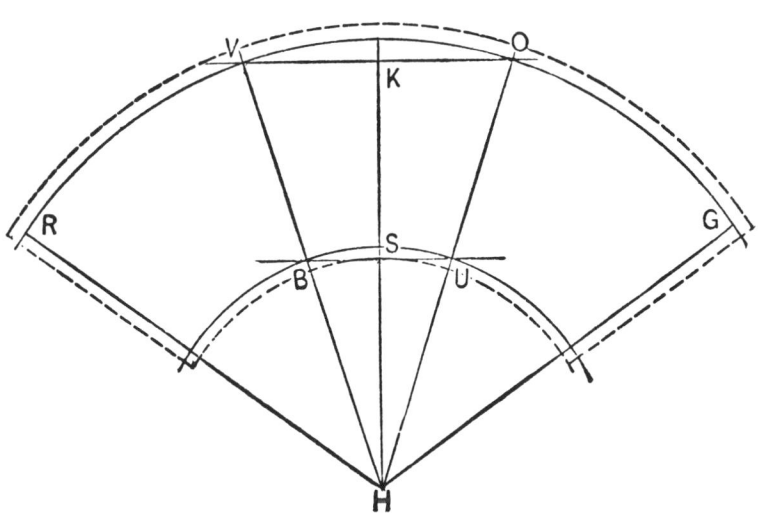

Draw perpendicular line H K, and from K lay off diameter of large end, as V O; on the line H K the hight of frustum, as K S. Draw line parallel to V O, and on it lay off small diameter, as B U. Draw lines through points V B and O U until they intersect at H. Span compasses from H to V and draw large arc R G; from H to B and describe small arc; make arc R G equal to circumference of large diameter and draw lines to center H. Allow for all edges, wire, burr and locks. This forms a pattern in one piece.

Rectangular Funnel.

Fig. 14.

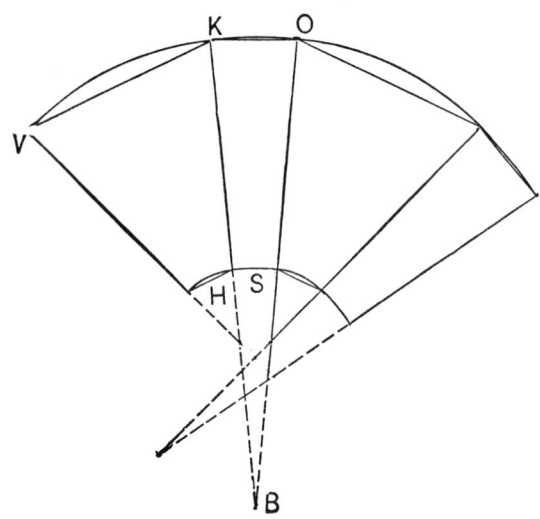

Draw side, as H K V. Continue side lines, as shown by dots. From point of intersection as center, describe arc and chord K V and H. Draw end O K S, producing lines to intersect at B. From B as center describe arc and chord O K and S. The other side and end obtained in the same manner, as shown in cut. Can be made in two or more pieces by dividing. All locks and edges must be allowed for on the pattern piece.

For Strainer Pail or Watering Pot Breast.

Fig. 15.

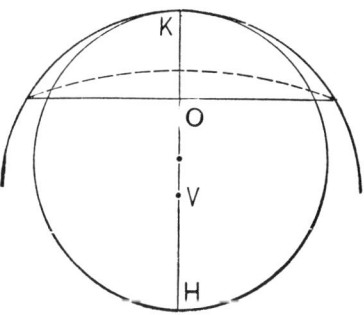

Strike circle size of pail or pot. Span dividers 1¾ inches, more or less, than radius of circle, being governed by pitch desired, as from V to K, and describe the arc. Draw the chord, making the segment K O which is the pattern of the desired width. The breast may be cut out if preferred, as shown by dotted lines.

Scale Tray or Scoop.

Fig. 16.

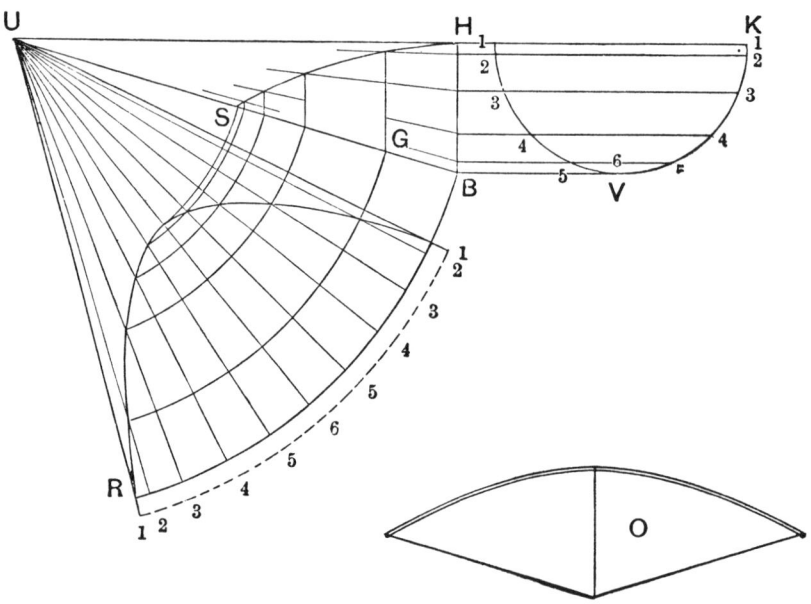

Construct a sectional view of the scoops, as H K V ; it being made in two pieces as O, let H S B represent one-half elevation of it. Continue the lines B S and K H until they cross at U. Divide H K V into any given number of spaces, continuing the same to the line H B, as shown by short lines. Draw lines from the division

points on H B to the joint U, thus obtaining the intersections on the line S H. With the T square at right
angles with H U, drop the points thus obtained on H S,
onto the line B S.

With U as center and U B as a radius describe the
arc B R. Step off upon this arc spaces equal to those in
H K V, using dividers, which gives the length B R.
Draw radial lines from U to space marks on line B R, as
shown.

With U as center and the various points on S B as
radii, describe arcs, intersecting similar radial lines as
shown. Then a line traced through the points thus
obtained, together with the arc B R, will be the outline
of the required pattern. Allow for edges, as shown by
dotted lines.

To Find Length of Sheet Required for Oval Boiler. Common Method.

Fig. 17.

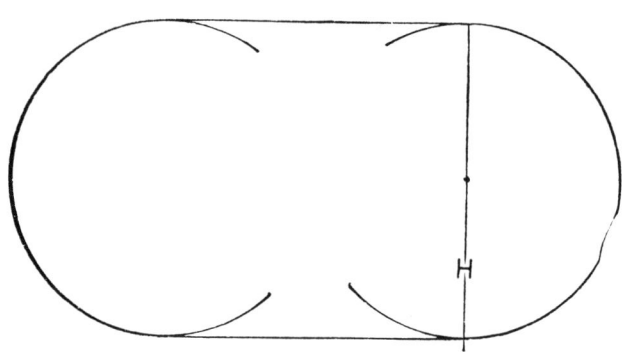

Describe bottom, length and width desired, then burr and from H as a starting point roll on the bench to obtain circumference. If three pieces are to be used, divide the circumference into three parts and allow edges; if made in two pieces, divide by two. Always divide the circumference by the number of pieces desired. Cut the cover the same size as bottom.

Oval Boiler Cover.

Fig. 18.

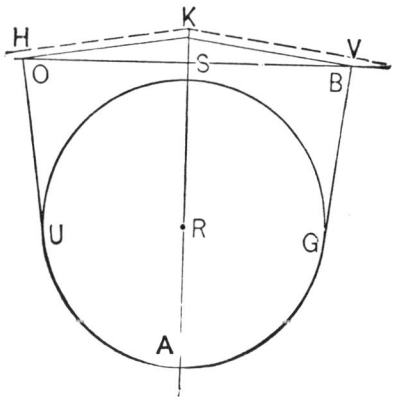

Draw line A K, and from R as center describe circle G U, size of boiler outside of rod. Make A K equal to one-half of entire length of boiler, and K S ⅜ inch or more if more pitch is desired. Through S draw the perpendicular line H V. Lay corner of square on line H, one blade at K, the other touching circle, describe lines U H K; in similar manner obtain K V G. Allow for locks and notch for edges.

Measure Lip.

Fig. 19.

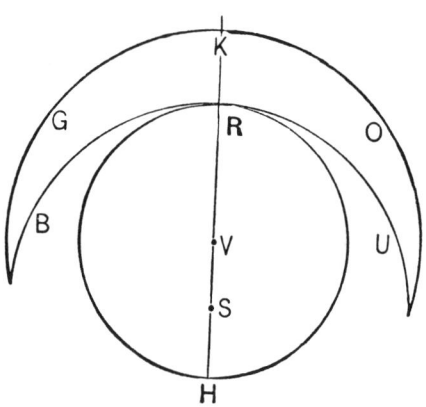

Draw line H K and upon it, with V as center, describe circle size of measure. With S as center, being the half distance from V to H, describe semicircle B. U. Make R K the desired width. With V as center describe G O. Cut on B U and G O to obtain the lip.

Steamer or Pitched Cover.

Fig. 20.

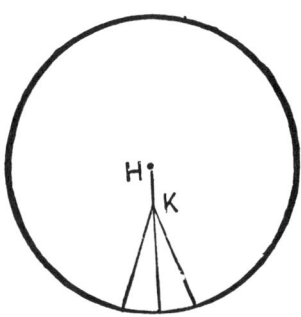

Strike circle 1 inch larger than rim burred. Draw line through center H, and from either side cut 1 inch on circle to 1 inch from center K. Draw lines and cut out. Or, strike circle the same or larger. Draw line through center and cut on it to center. After burring put in rim; draw up and mark, cut out triangular piece and solder. Much quicker and equally as good.

Heart with Square and Compass.

Fig. 21.

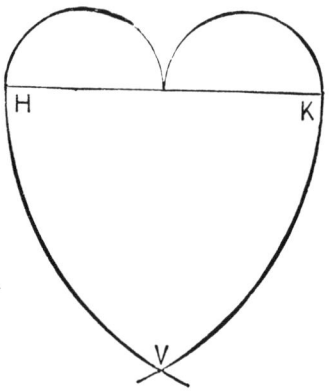

Draw line H K the breadth of the heart and on it two semicircles. Span dividers from H to K and make sweep to V.

To Describe a Star.

Fig. 22.

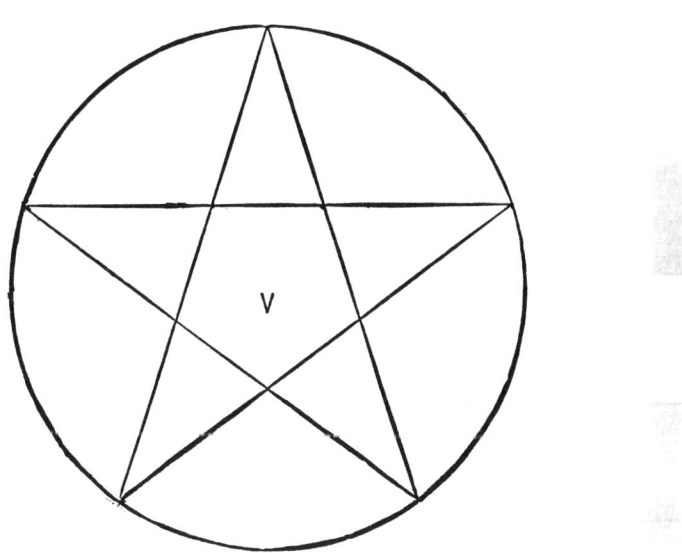

From V as center strike circle size of star desired. Divide circle in five parts and draw lines to points.

There is a rule for finding the points of a star other than stepping, but I do not give it. I have found the mode given to be the quickest and most accurate.

Pattern for Cutting Balls. —To Describe the Gores.

Fig. 23.

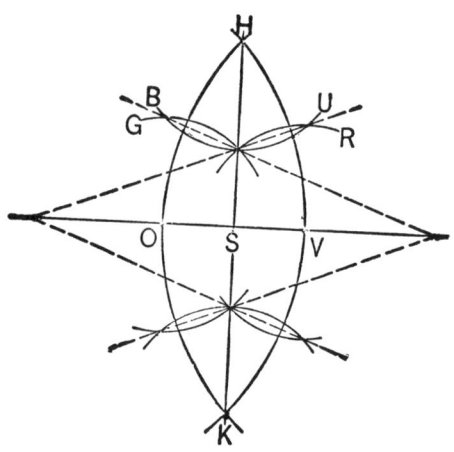

Erect perpendicular line H K equal to one-half the circumference of the ball; divide this line into one-half the
number of pieces required in full ball; make the line V O
equal to one of these pieces, cutting H K through the
center at right angles; then with H and K as centers, with
radius greater than one-half the distance K S, describe the
two arcs B U; with V and O as centers, arcs R G; draw

lines through these points, as shown by dotted lines. From points of intersection describe arcs H V K and H O K, and you obtain pattern for one piece. Allow for laps or seams. The more pieces used the better globe produced. Good results obtained by slightly raising the pieces.

To Describe an Oval.

Fig. 24.

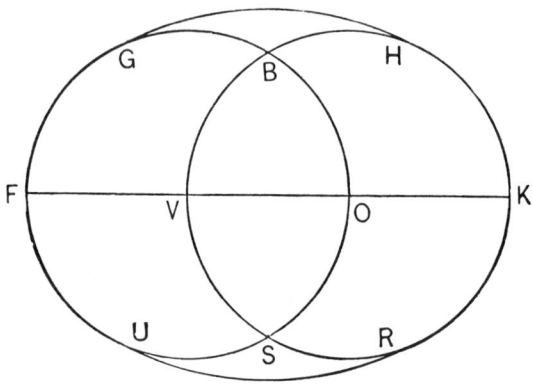

Draw horizontal line F K, span the dividers one-third the required major diameter, and from V and O as centers describe circles, as shown; then span dividers two-thirds entire length, and, with one foot at the intersection of the circles, as S and B, draw the arcs G H and U R, which completes the oval.

The proportion of the diameters is about as 3 to 4.

To Describe Oval with Diameters as 5 to 8.

Fig. 25.

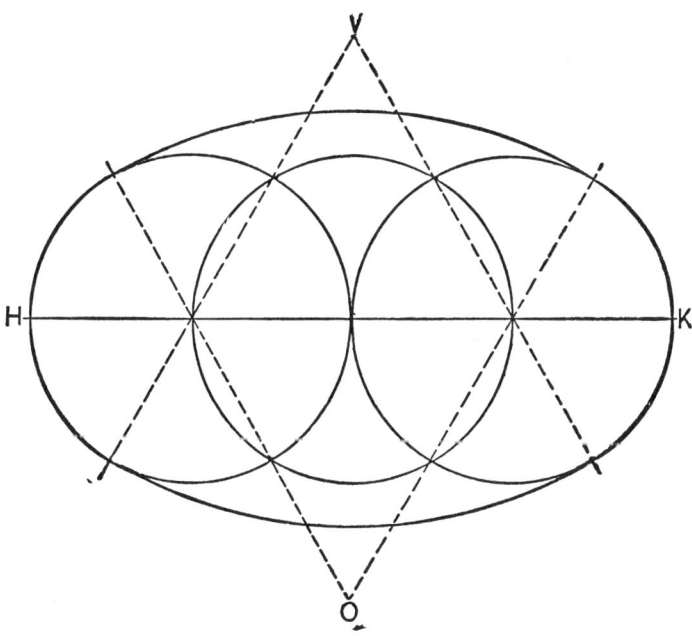

Draw horizontal line H K. Span compasses one-quarter the long diameter and describe three circles with that radius, as shown by diagram. Then draw lines through centers of outer circles and their intersections, as shown. The oval is completed by drawing the arcs connecting the outer circles from points V and O as centers.

To Describe an Oval. Another Method.

Fig. 26.

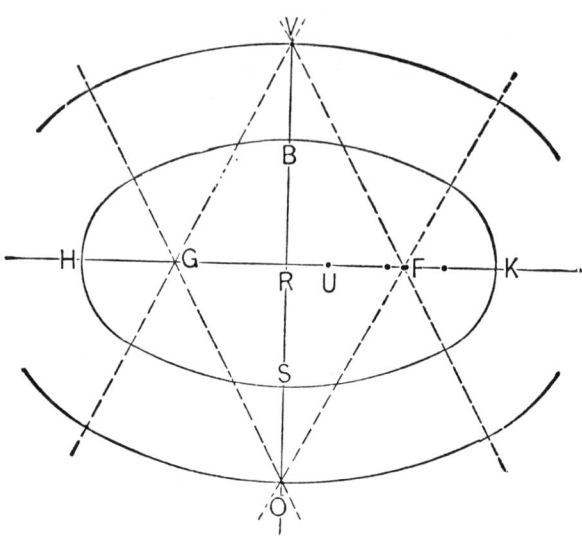

Draw horizontal line H K and perpendicular to it V C. Let H K equal the long or transverse diameter, and S B the short or conjugate. Lay off the distance S B on the line H K, as from H to U. Divide the distance U K into three equal parts. From R, the center, set off two of the parts each side, as G F. On the line V O set off the distance G F from R, as R V and R O. From V and O draw lines passing through G and F, as shown. From the points V, O, G, F as centers describe the arcs that complete the ellipse.

To Describe an Oval. Another Method.

Fig. 27.

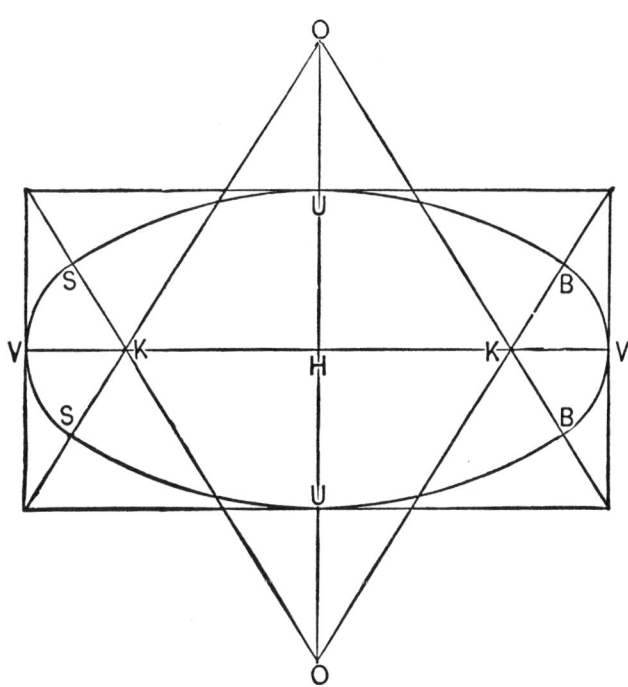

Construct the parallelogram equal in length and width to the long and short diameters of the oval desired. Divide it into four equal parts by drawing lines through the center, crossing at H. Mark the points K and K one-third the distance from V to H, and draw lines from the corners through these points until they intersect, as shown at O. Then from O and O as centers describe the arcs S U B and S U B; from K and K as centers the segments B V B and S V S.

To Describe Oval by Means of String, Pins and Pencil.

Fig. 28.

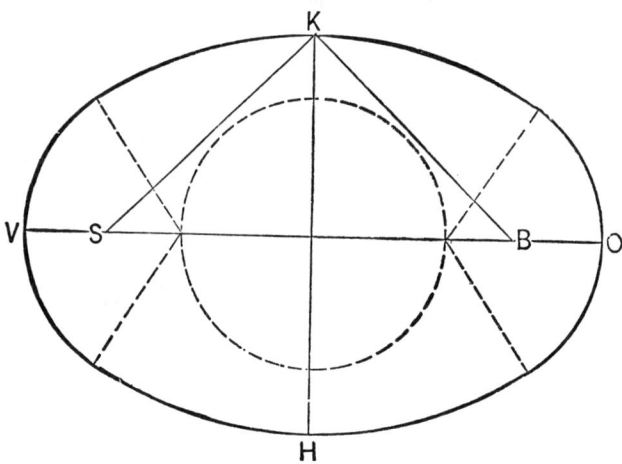

Erect perpendicular line H K equal to short diameter and at right angles to it V O. Span dividers one-half the length of the oval, and with H and K as centers describe the arcs S and B. Set pins at these points, and, with a string (one that will not stretch) tied around them so that the loop when drawn tight will reach H or K, as shown, draw the figure with pencil, keeping string equally tense while going around. Of all the apparatus invented for oval drawing I think the string is the best. The best results, at least, are obtained. To attempt to draw a perfect oval or ellipse by the use of compasses is vain. It cannot be done so that the line will be true, or the proportion or shape satisfactory to one with an eye for correctness or uniformity. The so-called trammels are the next best thing, but no better. A few rules for drawing ovals by the use of dividers have been given in this work so the mechanic may take his choice, and after a little practice with the string and nails will find them the best trammels yet invented for the purpose.

To Describe Pattern for Flaring Oval Vessel.
Two Pieces.

Fig. 29.

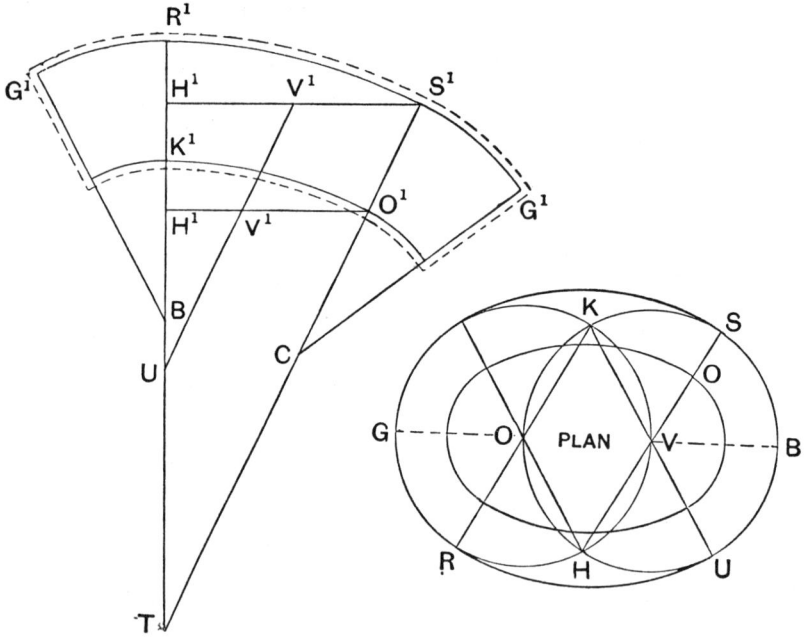

Draw plan according to rule given in Fig. 24, or any other method. Construct right angle triangle T H¹ S¹ and parallel to H¹ S¹, draw H¹ O¹, the distance between hight of article. Lay off on H¹ S¹ the distances H S and V S in plan and on H¹ O¹ the distances H O and V O in plan. Draw lines through these points to intersect the line R^1 T at U and T. Using T as center draw the arcs O^1 K^1 and S^1 R^1, making the distance along the arc S^1 R^1 equal to U R in plan. Draw line from R^1 to T. Take radius V^1 U on the lines R^1 T and S^1 T and obtain centers B and C, with which describe the arcs R^1 G^1 and S^1 G^1, which make equal in length to G R or U B in plan. Draw lines to centers B and C. Allow for all edges, locks, wire and burr.

To Describe Pattern for Flaring Article with Straight Sides and Round Ends.   Two Pieces.

Fig. 30.

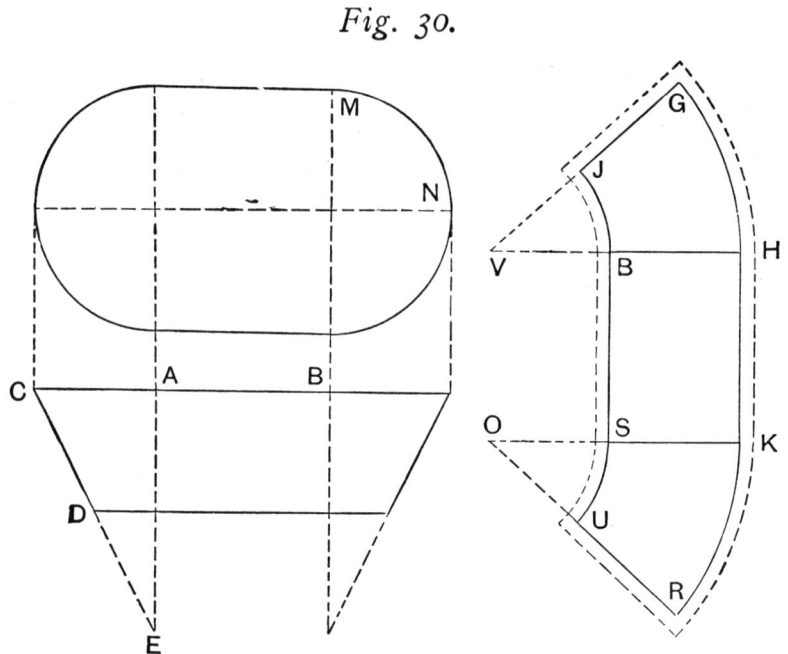

Erect two perpendicular lines, H V, K O, distance between the length of sides A B; at right angles to these, two lines, distance between the slant hight of article C D. On H V and K O set off the radius C E as V and O. From V and O as centers, with radii V B, V H and O S, O K, draw the arcs B J, H G and S U, K R.   Make the arcs H G and K R equal to one-half the circumference of the ends M N and draw lines to V and O.   Allow for all edges, locks, wire and burr.

To Describe Pattern for Oval Flaring Vessel. Four Pieces.

Fig. 31.

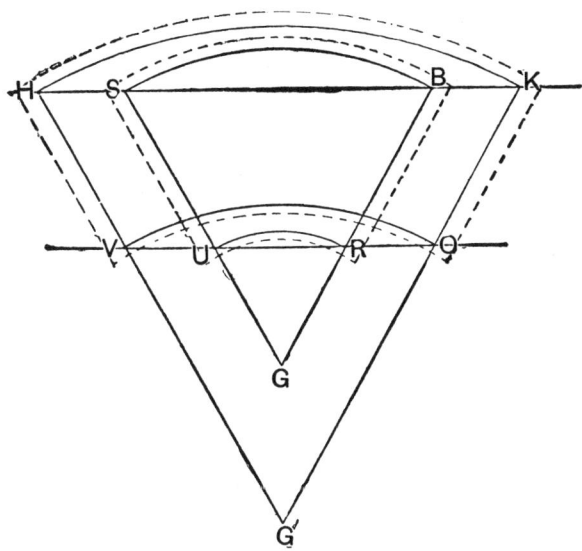

Describe bottom as by Fig. 27. Obtain length of arcs S U B and S V S, also length of corresponding arcs at the top of vessel. Draw horizontal lines H K and V O, making the distance between the desired slant hight. Make H K equal in length to that of the piece at the top, and V O to that of the bottom, for the sides. S B and U R for the end pieces. Produce lines through these points to intersect at G and G'. Describe the arcs from these points. Allow for all edges, locks, wire and burr.

To Describe Pattern for Flaring Hexagon Article.

Fig. 32.

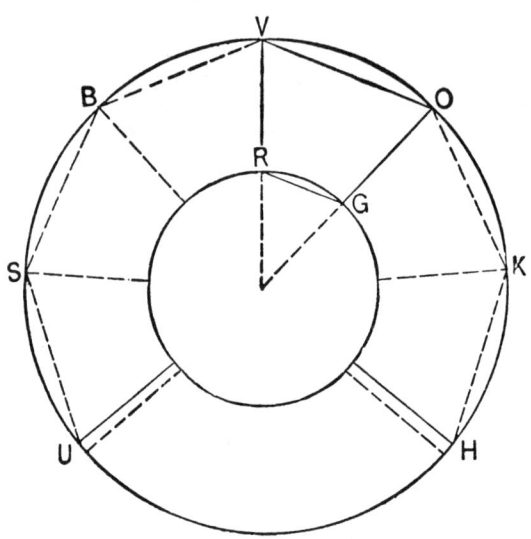

Let V O represent width of the bottom of one side and R G the width of the top of one side, the distance between the slant hight. Produce side lines until they cross in the center, as shown by dotted lines. Span dividers from center to O, and describe circle H O U; span to G and describe inner circle; span again from V to O and step on the outer circle three spaces each side from O, as K, H, B, S, U. Draw lines from these points tending toward center, and connect by chords as H K, K O, etc. Cut out piece H U, allowing for locks, as shown. Pattern for a pentagon article may be described by the same rule.

To Describe Pattern for Flaring Square Vessel.

Fig. 33.

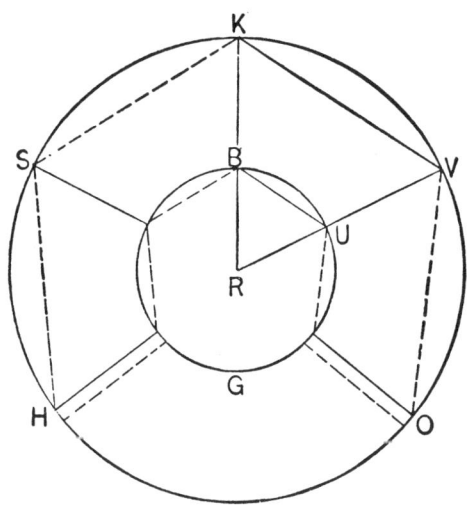

Let K V and B U represent the width of the bottom and top of one of the sides, the distance between the slant hight. Continue lines until they intersect at R. With radius R B. strike circle U B G. Span dividers from K to V and set off on outer circle the distance, as V O, K S, etc.; draw lines through these points tending toward the center R, also the chords, as shown by dotted lines. Allow for edges. Can be made in two pieces by dividing and allowing for extra lock or seam.

To Describe Pattern for Flaring Article with Square Top and Base a Rectangle. Two or Four Pieces.

Fig. 34.

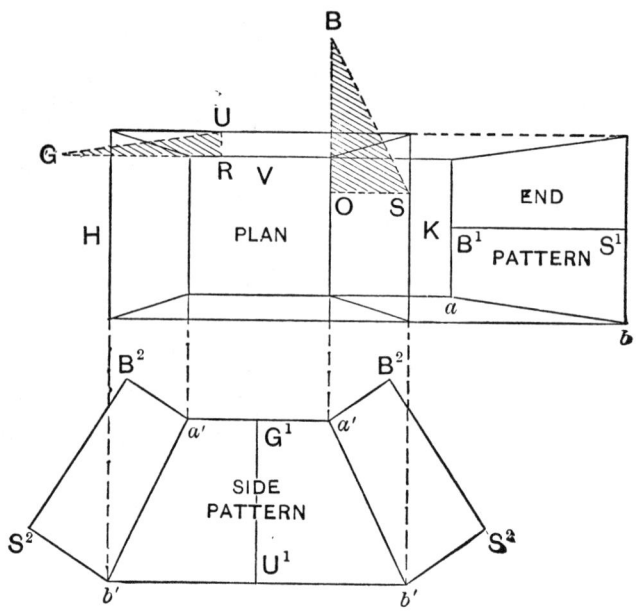

Draw rectangular base H K and square top V in center of base. Draw perpendiculars O S and R U. Also place the hight of the article O B and R G. Place the slant hight B S on $B^1 S^1$ and draw lines a and b which intersect as shown, which gives pattern for end. Place G U on $G^1 U^1$, draw lines a' and b' which intersect as shown, which gives pattern for side. Join half of end pattern to either side of side pattern as shown by similar letters, which gives half pattern.

To Describe Tapering Octagon Article.

Fig. 35.

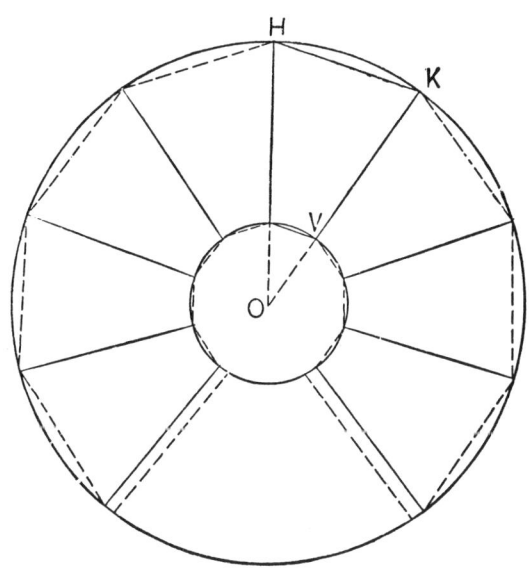

Draw bottom K H and top V of one side, with distance between the slant hight, and continue side lines until
they intersect at O. With O as a center and the radii
O V and O H, describe inner and outer circles. Set off
on them distances equal to H K and V, and connect by
chords, as shown by dotted lines. Allow for locks and
edges.

Flaring Article, Top and Base a Rectangle. Two Pieces.

Fig. 36.

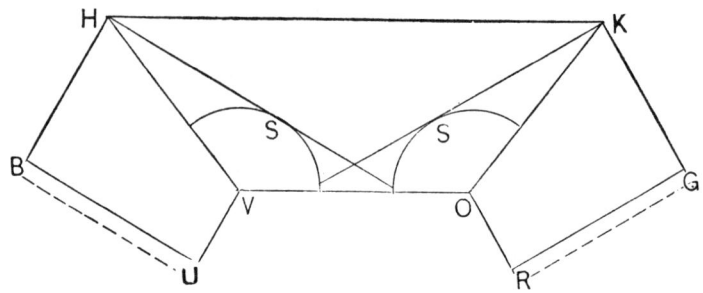

Draw side elevation, as H K, V O, of the longest side. Span dividers the difference between the shortest side of the base and longest side of top. From V and O as centers describe arcs S and S. With blade of square resting on arcs and the corner at H and K, draw lines H B and K G. Set off H B and K G equal one-half of shortest sides of base and draw lines B U and G R at right angles to H B and K G; also lines U V and R O at right angles to U B and G R. Allow for locks, as shown by dotted lines. For a strictly accurate pattern proceed as in Fig. 34.

Round Base and Square Top Article. Two Pieces.

Fig. 37.

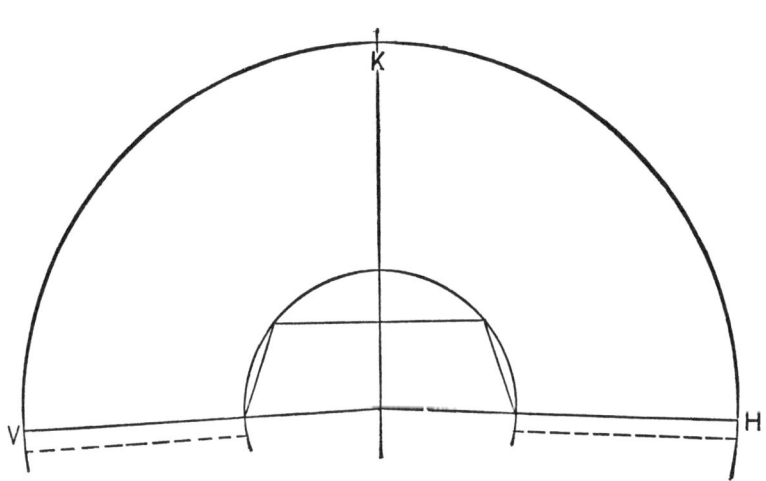

Erect perpendicular line. Span dividers to three-quarters diameter of base and describe semicircle H K V. Make K V and K H each equal to one-quarter the circumference of the round base and draw lines to center. Span dividers to three-quarters size of top from corner to corner and describe inner circle. Lay out sides of top, size required, on circle, as shown. Allow laps.

Rectangular Base and Round Top Article.
Two Pieces.

Fig. 38.

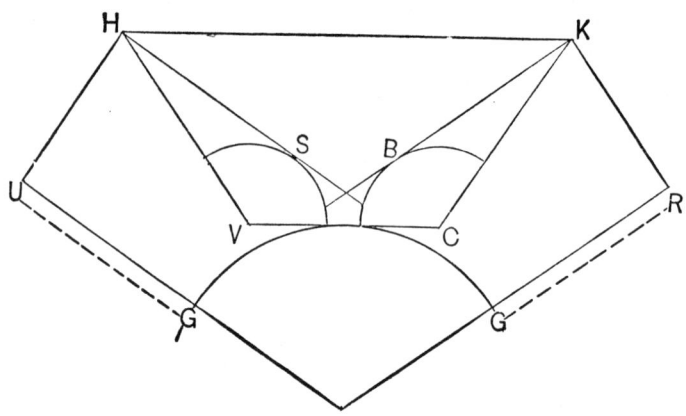

Draw horizontal lines H K, V O. Make H K equal to the longest side of base, V O equal to one-fourth the circumference of the top, the distance between slant hight; draw side lines through these points. With radii one-half the difference between V O and the shortest side of the base, describe the arcs S, B; with blade of square resting on arcs, and corner at H and K, draw lines K R, H U, equal to one-half the short side; at right angles to K R, H U, draw lines R G and U G; U G and R G produced will intersect; from this point span dividers to line V O and describe the arc. Allow for locks and edges.

Square Base and Round Top Article. Two Pieces.

Fig. 39.

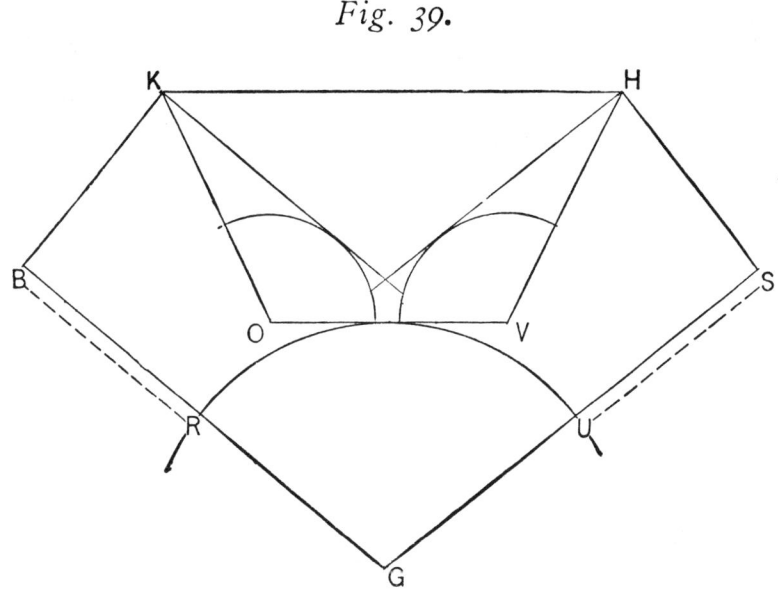

Draw horizontal lines H K, V O; H K equal to the length of one side of the base, V O equal to one-fourth the circumference of the top, the distance between the slant hight; draw lines through these points. With radii one-half the difference between K H and O V, describe arcs; with blade of square resting on arcs and the corner at H and K, draw lines H S and K B, equal to one-half the base; at right angles to H S and K B draw S U and B R, produced to intersect at G. Span dividers from G to line V O and describe the arc. Allow for locks and edges.

To Describe a Square or Right Angle Elbow.
Two Pieces.

Fig. 40.

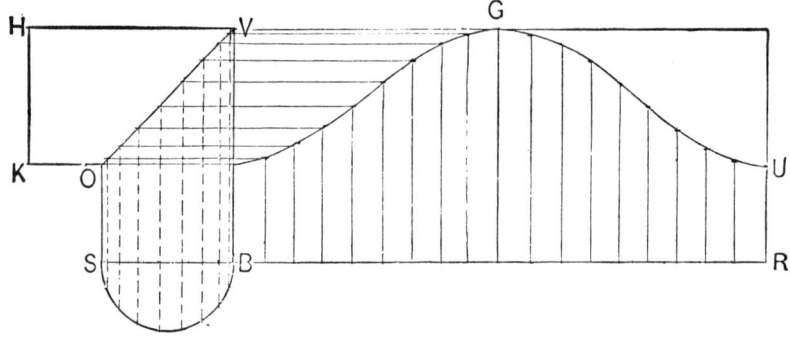

Draw the elevation of the elbow, as B S, O V, K H. Draw line from V to O. Divide one-half of the plan into a convenient number of equal parts, as shown by dotted lines; erect lines to intersect O V. Make the line B R equal in length to the circumference of the elbow. Set off on this line spaces corresponding to those in the plan, the same number each side of the center line; then draw lines parallel to the arm of the elbow, cutting the corresponding lines as indicated. By tracing through these points the irregular line U G the pattern is obtained. Allow for locks or rivets.

The general principle for cutting elbow patterns is the same throughout, and to understand the principle is to be able to describe pattern for any elbow, at any angle and of any number of pieces. It is the design of this work to make the principle clear.

Quick Method.

Fig. 41.

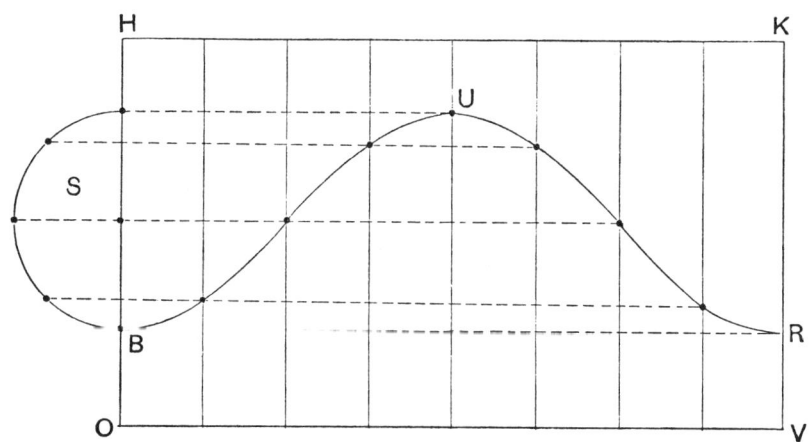

Lay out on sheet length required for elbow, as H K V O. Describe semicircle S the desired size of pipe, which divide into four parts. Space the length of the sheet into twice the number of squares in S, and draw vertical and horizontal lines until they intersect. O B U R V is then an accurate pattern. Allow for flanges.

To Describe Three-Piece Elbow.

Fig. 42.

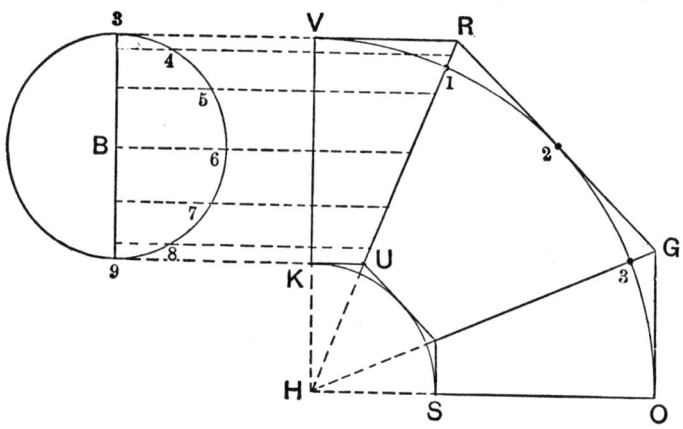

Let H K be the throat and K V the diameter of the elbow. Draw the quadrant V O, which divide into four equal parts, as shown by 1, 2, 3. Draw miter lines through 1 and 3 as H R and H G. Draw the circle B equal to diameter of elbow and divide one-half of B in equal parts, as shown; draw lines to intersect miter line R U.

Fig. 43.

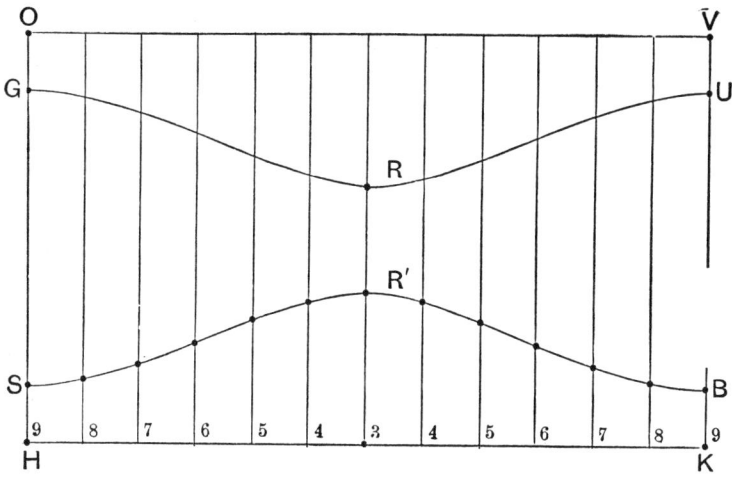

Construct parallelogram H K V O equal in length to the circumference of B. Through the spaces on H K draw parallel lines as shown. Measuring from V K, take the various distances to the miter line R U and place them on similar lines measuring from H K. H S B K is then the pattern for the end. Double the distance from 3 to R[1] and place it from S to G and B to U and transfer the miter line S R[1] B to G R U. Place H S as shown by G O and U V and draw O V, which completes the three patterns.

To Describe a Right Angle Elbow.   Four Pieces.

Fig. 44.

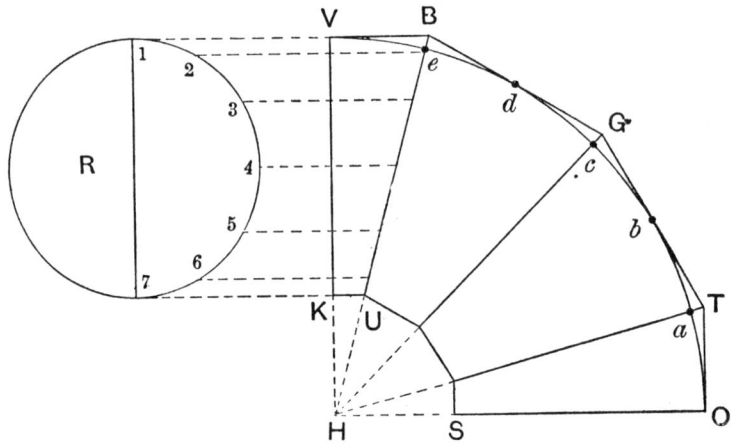

Let H K be the throat and K V the diameter of the elbow. Draw the quarter circle V O, which divide into six equal parts, as shown by a b c d e. Draw miter lines through a, c and e, as shown by H B, H G and H T. Draw the circle R, which space as shown, and draw lines to intersect the miter line B U.

Fig. 45.

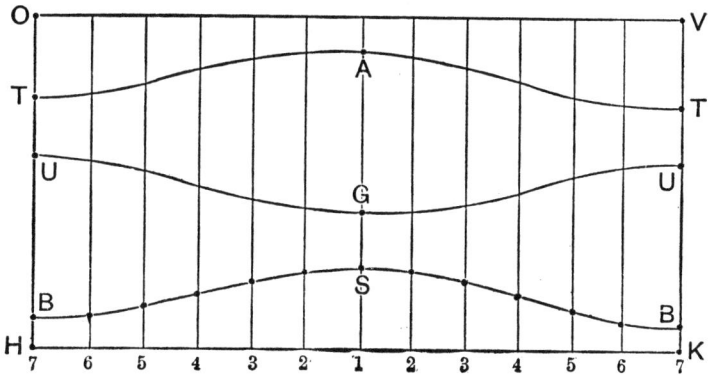

Construct parallelogram H K V O, equal in length to the circle R, as shown by similar figures on H K, through which draw parallel lines as shown. Measuring from V K, take the various distances to the miter line B U and place them on similar lines in the pattern, measuring from H K, and obtain B S B. Double 1 S and place at B U and B U and trace the miter cut B S B as shown by U G U. Place S G at U T and U T and trace U G U as shown by T A T. Make T O and T V equal to S 1 and draw line O V, which completes the four patterns. Allow for locks.

Elbow in Five Sections.

Fig. 46.

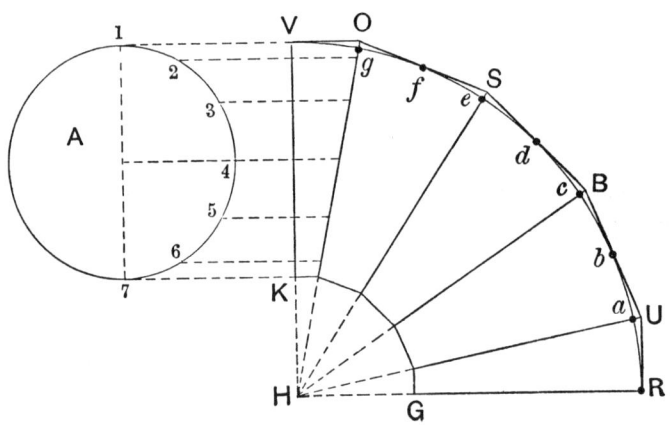

Draw throat H K and diameter K V. Draw quadrant H V R, which divide into eight parts as shown from a to g; draw miter lines H U, H B, H S and H O. Divide profile A into equal spaces, and draw lines to miter line H O.

Fig. 47.

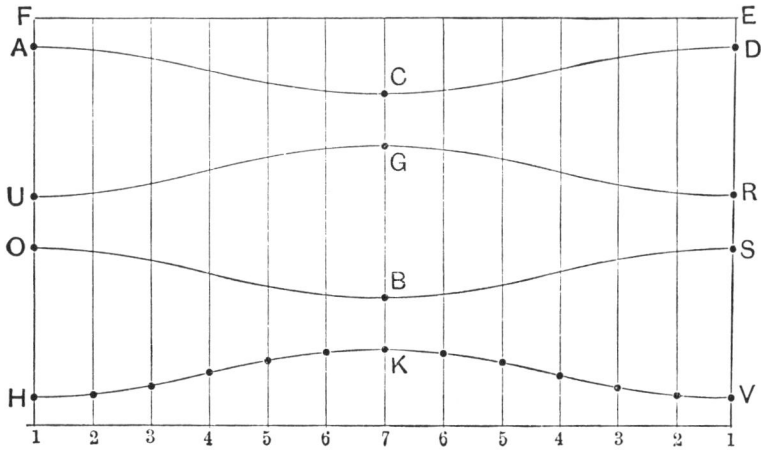

Make 1 1 equal to circumference of profile A. Draw parallel lines as shown in pattern. Use dividers and measure various distances from V K to miter line H O, which transfer to similar lines measuring from 1 1, and obtain miter cut H K V. Double 7 K and place at H O and V S and draw miter cut O B S. Place K B at O U and S R and draw miter cut U G R. Make U A and R D equal to H O and draw miter cut A C D. Make A F and D E equal to H 1 and draw F E, which completes the five patterns. Allow for locks.

To Describe Pattern for Obtuse Elbow.

Fig. 48.

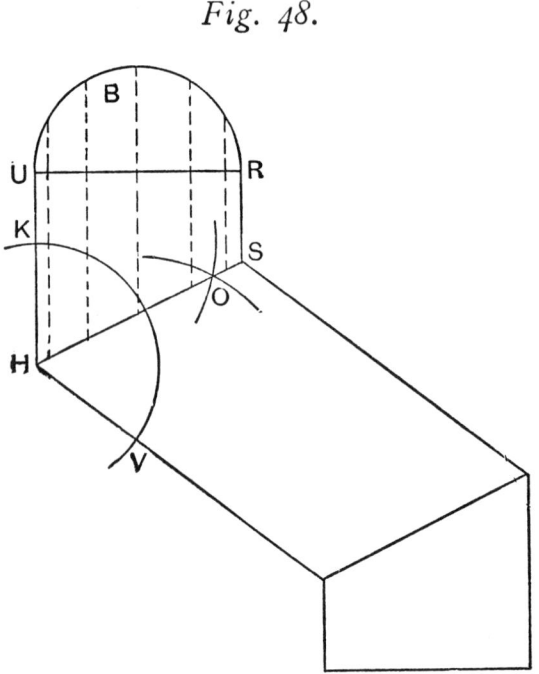

When the pattern for an obtuse elbow is desired it is only necessary to draw a correct representation of the elbow and obtain the miter line, as follows: With H as center, draw the arc K V. With any desired radius, and using K and V as centers, intersect arcs at O. Draw the miter line H O S. Place the half profile B in position as shown, which space, and draw parallel lines to the miter line H S. Then proceed as by the rules already given, and the result will be satisfactory.

To Describe a Tapering Elbow

Fig. 49.

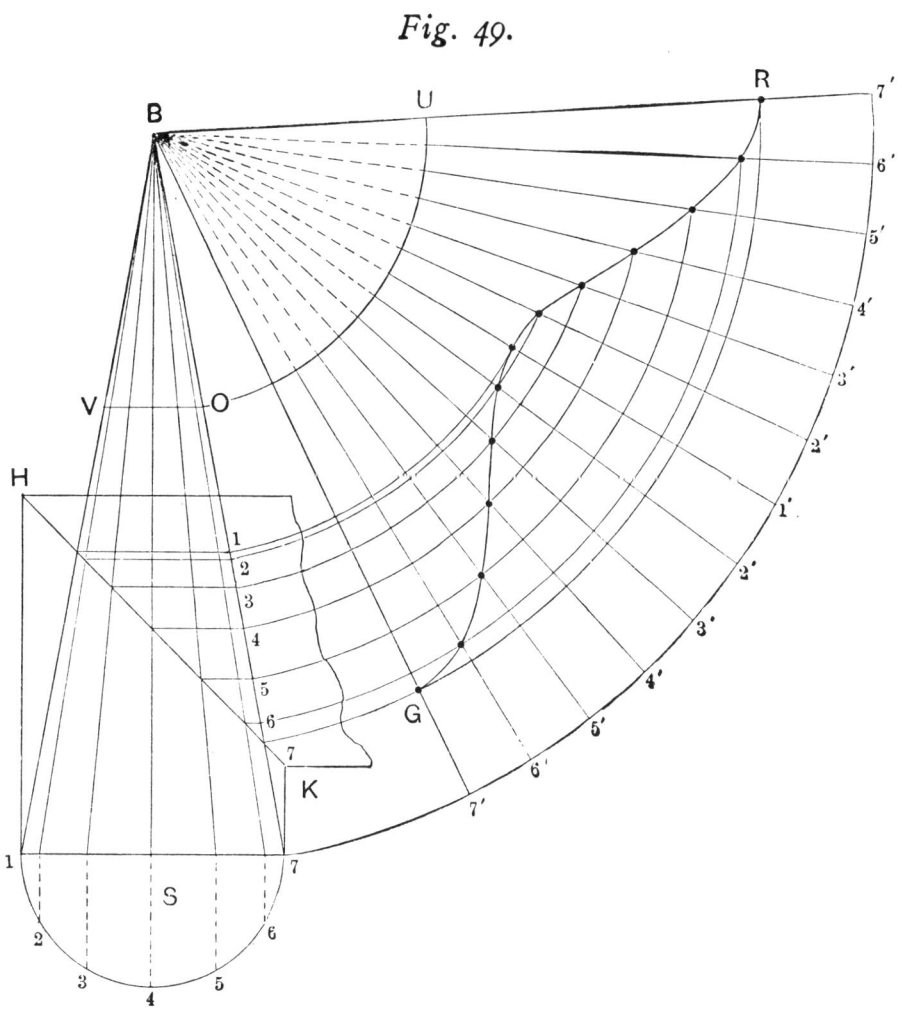

Draw elevation of elbow at any angle desired and draw miter line H K as shown. Establish hight and diameter of small end as V O and extend the lines 1-V and 7-O until they meet at B. Draw half profile S, which space into equal parts and draw vertical lines to 1-7, from which draw radial lines to the apex B, which will cross the miter line H K as shown. From these intersections draw horizontal lines to the side B-7 as shown from 1 to 7. With B-7 as radius, draw the arc 7'-7' equal to the circumference of the circle S. From the points on 7'-7' draw radial lines to the apex B, which intersect by arcs struck from B as center, with radii equal to the points between 1 and 7. U R G O is the pattern for the upper arm and R G 7'-7' pattern for the lower arm. Allow for locks.

To Obtain Length of Piece for Tea Kettle Body.

Fig. 50.

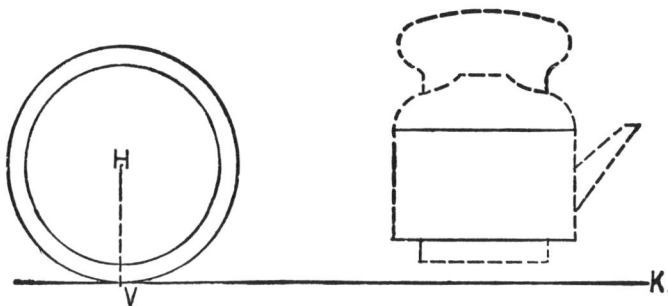

The way in general practice is to roll the bottom after burring on the bench to obtain circumference, and use strip ¾ inch less in length, as shown by figure. H represents the pit; K V the length of the strip or sheet.

Mode of Stringing Patterns.

Fig. 51.

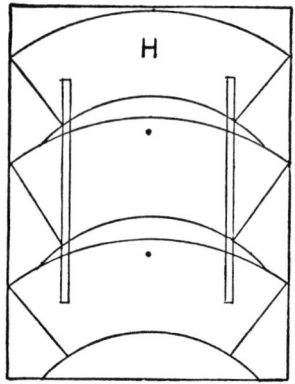

This cut represents the three pieces of a 6-quart pan usually cut from one sheet of 10 x 14 tin. Instead of using one piece for pattern and placing it three times, three pieces are fastened together by soldering on two strips of tin with a heavy hem on each side, and all placed at once, thus saving time and vexation. To use to advantage begin at the bottom of the string pattern and mark around on the outside first, and then mark in the centers.

String Pattern.

Fig. 52.

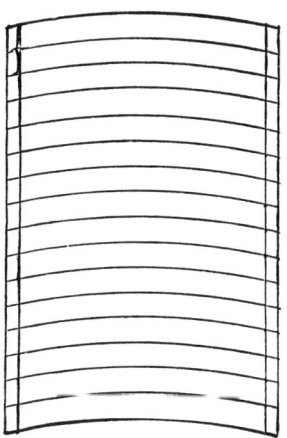

This figure represents a string of rim or hoop patterns, fastened as shown in the same manner as described on page 64. Rims of any width can be put together in this manner and a great saving of time is the result when once properly done. Patterns for all articles of tinware should be strung in this way, when more than one piece is obtained from a sheet, that the marking out may be expedited and less tedious.

Description of Boiler Block.

Fig. 53.

By this figure is represented a block for truing up boilers after they are formed up in the rollers and locked together. Many mechanics depend upon the stake and the accuracy of the eye, but after using this method would not abandon it, as better results are obtained and in much less time. The block is made of 2-inch plank, by placing one on another and securing with four long bolts passing through them. The proper dimensions are as follows:

Bottom, 13 inches wide, 25 inches long.

Top,   10   "   "   19   "   "

Hight,   12   "

APPENDIX.

EPITOME OF MENSURATION.

OF THE CIRCLE, CYLINDER, SPHERE, ETC.

1. The circle contains a greater area than any other plane figure bounded by an equal perimeter or outline.

2. The areas of circles are to each other as the squares of their diameters.

3. The diameter of a circle being 1, its circumference equals 3.1416.

4. The diameter of a circle is equal to .31831 of its circumference.

5. The square of the diameter of a circle being 1, its area equals .7854.

6. The square root of the area of a circle multiplied by 1.12837 equals its diameter.

7. The diameter of a circle multiplied by .8862, or the circumference multiplied by .2821, equals the side of a square of equal area.

8. The number of degrees contained in the arc of a circle multiplied by the diameter of the circle and by .008727, the product equals the length of the arc in equal terms of unity.

9. The length of the arc of a sector of a circle multiplied by its radius equals twice the area of the sector.

10. The area of the segment of a circle equals the area of the sector, minus the area of a triangle whose vertex

is the center and whose base equals the chord of the segment.

11. The sum of the diameters of two concentric circles multiplied by their difference and by .7854 equals the area of the ring or space contained between them.

12. The circumference of a cylinder multiplied by its length or hight equals its convex surface.

13. The area of the end of a clyinder multiplied by its length equals its solid contents.

14. The area of the internal diameter of a cylinder multiplied by its depth equals its cubical capacity.

15. The square of the diameter of a cylinder multiplied by its length and divided by any other required length, the square root of the quotient equals the diameter of the other cylinder of equal contents or capacity.

16. The square of the diameter of a sphere multiplied by 3.1416 equals its convex surface.

17. The cube of the diameter of a sphere multiplied by .5236 equals its solid contents.

18. The hight of any spherical segment or zone, multiplied by the diameter of the sphere of which it is a part and by 3.1416, equals the area or convex surface of the segment; òr,

19. The hight of the segment multiplied by the circumference of the sphere of which it is a part equals the area.

20. The solidity of any spherical segment is equal to three times the square of the radius of its base, plus the square of its hight, multiplied by its hight and by .5236.

21. The solidity of a spherical zone equals the sum of the squares of the radii of its two ends and one-third

the square of its hight, multiplied by the hight and by 1.5708.

22. The capacity of a cylinder, 1 foot in diameter and 1 foot in length, equals 5.875 United States gallons.

23. The capacity of a cylinder, 1 inch in diameter and 1 foot in length, equals .0408 United States gallon.

24. The capacity of a cylinder, 1 inch in diameter and 1 inch in length, equals .0034 United States gallon.

25. The capacity of a sphere 1 foot in diameter equals 3.9168 United States gallons.

26. The capacity of a sphere 1 inch in diameter equals .002266 United States gallon ; hence,

27. The capacity of any other cylinder in United States gallons is obtained by multiplying the square of its diameter by its length, or the capacity of any other sphere by the cube of its diameter and by the number of United States gallons contained as above in the unity of its measurement.

OF THE SQUARE, RECTANGLE, CUBE, ETC.

1. The side of a square equals the square root of its area.

2. The area of a square equals the square of one of its sides.

3. The diagonal of a square equals the square root of twice the square of its side.

4. The side of a square is equal to the square root of half the square of its diagonal.

5. The side of a square equal to the diagonal of a given square contains double the area of the given square.

6. The area of a rectangle equals its length multiplied by its breadth.

7. The length of a recangle equals the area divided by the breadth; or the breadth equals the area divided by the length.

8. The solidity of a cube equals the area of one of its sides multiplied by the length or breadth of one of its sides.

9. The length of a side of a cube equals the cube root of its solidity.

10. The capacity of a 12-inch tube equals 7.48 United States gallons.

OF TRIANGLES, POLYGONS, ETC.

1. The complement of an angle is its defect from a right angle.

2. The supplement of an angle is its defect from two right angles.

3. The three angles of every triangle are equal to two right angles: hence the oblique angles of a right angled triangle are each other's complements.

4. The sum of the squares of two given sides of a right angled triangle is equal to the square of the hypothenuse.

5. The difference between the squares of the hypothenuse and given side of a right angled triangle is equal to the square of the required side.

6. The area of a triangle equals half the product of the base multiplied by the perpendicular hight.

7. The side of any regular polygon multiplied by its apothem or perpendicular, and by the number of its sides, equals twice the area.

OF ELLIPSES, CONES, FRUSTUMS, ETC.

1. The square root of half the sum of the squares of the two diameters of an ellipse multiplied by 3.1416 equals its circumference.

2. The product of the two axes of an ellipse multiplied by .7854 equals its area.

3. The curve surface of a cone is equal to half the product of the circumference of its base multiplied by its slant side, to which, if the area of the base be added, the sum is the whole surface.

4. The solidity of a cone equals one-third the product of its base multiplied by its altitude or hight.

5. The square of the diameters of the two ends of the frustum of a cone added to the product of the two diameters, and that sum multiplied by its hight and by .2618, equals its solidity.

DEFINITIONS OF ARITHMETICAL SIGNS USED IN THE FOLLOWING CALCULATIONS.

$=$ Sign of Equality, and signifies as $4 + 6 = 10$.

$+$ " Addition, " as $6 + 6 = 12$, the Sum

$-$ " Subtraction, " as $6 - 2 = 4$, Remainder.

$\times$ " Multiplication, " as $8 \times 3 = 24$, Product.

$\div$ " Division, " as $24 \div 3 = 8$,

$\sqrt{}$ " Square Root, " Extraction of Square Root.

6^2 " to be squared, " thus $8^2 = 64$.

7^3 " to be cubed, " thus $3^3 = 27$.

DECIMAL EQUIVALENTS TO FRACTIONAL PARTS OF LINEAL MEASUREMENT.

ONE INCH THE INTEGER OR WHOLE NUMBER.

.96875	equal	⅞ and 3-32	.46875	equal	⅜ and 3-32
.9375	"	⅞ and 1-16	.4375	"	⅜ and 1-16
.90625	"	⅞ and 1-32	.40625	"	⅜ and 1-32
.875	"	⅞	.375	"	⅜
.84375	"	¾ and 3-32	.34375	"	¼ and 3-32
.8125	"	¾ and 1-16	.3125	"	¼ and 1-16
.78125	"	¾ and 1-32	.28125	"	¼ and 1-32
.75	"	¾	.25	"	¼
.71875	"	⅝ and 3-32	.21875	"	⅛ and 3-32
.6875	"	⅝ and 1-16	.1875	"	⅛ and 1-16
.65625	"	⅝ and 1-32	.15625	"	⅛ and 1-32
.625	"	⅝	.125	"	⅛
.59375	"	½ and 3-32	.09375	"	3-32
.5625	"	½ and 1-16	.0625	"	1-16
.53125	"	½ and 1-32	.03125	"	1-32
.5	"	½			

ONE FOOT OR TWELVE INCHES THE INTEGER.

.9166	equal	11 inches.	.1666	equal	2 inches.
.8333	"	10 "	.0833	"	1 "
.75	"	9 "	.07291	"	⅞ "
.6666	"	8 "	.0625	"	¾ "
.5833	"	7 "	.05208	"	⅝ "
.5	"	6 "	.04166	"	½ "
.4166	"	5 "	.03125	"	⅜ "
.3333	"	4 "	.02083	"	¼ "
.25	"	3 "	.01041	"	⅛ "

MENSURATION OF SURFACES.

MENSURATION is that branch of Mathematics which is employed in ascertaining the extension, solidities and capacities of bodies capable of being measured.

MENSURATION OF SURFACES.

To Measure or Ascertain the Quantity of Surface in Any Right Lined Figure whose Sides are Parallel to Each Other.

RULE.—*Multiply the length by the breadth or perpendicular hight, and the product will be the area or superficial contents.*

APPLICATION OF THE RULE TO PRACTICAL PURPOSES.

The sides of a square piece of iron are 9⅞ inches in length, required the area.

Decimal equivalent to the fraction ⅞ = .875, and 9.875 × 9.875 = 97.5, etc., square inches, the area.

The length of a roof is 60 feet 4 inches and its width 25 feet 3 inches; required the area of the roof.

4 inches = .333 and 3 inches = .25 (see table of equivalents), hence, 60.333 × 25.25 = 1523.4 square feet, the area.

TRIANGLES.

To Find the Area of a Triangle When the Base and Perpendicular are Given.

RULE.—*Multiply the base by the perpendicular hight and half the product is the area.*

The base of the triangle is 3 feet 6 inches in length and the hight 1 foot 9 inches; required the area.

6 in. = .5 and 9 in. = .75, hence, $\dfrac{3.5 \times 1.75}{2} = 3.0625$

square feet, the area.

Any Two Sides of a Right Angled Triangle being Given, to Find the Third.

WHEN THE BASE AND PERPENDICULAR ARE GIVEN TO FIND THE HYPOTHENUSE.

Add the square of the base to the square of the perpendicular and the square root of the sum will be the hypothenuse.

The base of the triangle is 4 feet and the perpendicular 3 feet; then $4^2 + 3^2 = 25$, $\sqrt{25} = 5$ feet, the hypothenuse.

WHEN THE HYPOTHENUSE AND BASE ARE GIVEN TO FIND THE PERPENDICULAR.

From the square of the hypothenuse subtract the square of the base, and the square root of the remainder will be the perpendicular.

The hypothenuse of the triangle is 5 feet and the base 4 feet; then $5^2 - 4^2 = 9$, and $\sqrt{9} = 3$, the perpendicular.

WHEN THE HYPOTHENUSE AND PERPENDICULAR ARE GIVEN TO FIND THE BASE.

From the square of the hypothenuse subtract the square of the perpendicular, and the square root of the remainder will be the base.

OF POLYGONS.

To Find the Area of a Regular Polygon.

RULE.—*Multiply the length of a side by half the distance from the side to the center, and that product by the number of sides; the last product will be the area of the figure.*

EXAMPLE.—The side of a regular hexagon in 12 inches, and the distance therefrom to the center of the figure is 10 inches; required the area of the hexagon.

$$\frac{10}{2} \times 12 \times 6 = 360 \text{ square inches} = 2\frac{1}{2} \text{ square feet. Ans.}$$

To Find the Area of a Regular Polygon when the Side Only is Given.

RULE.—*Multiply the square of the side by the multiplier opposite to the name of the polygon in the ninth column of the following table, and the product will be the area.*

Table of angles relative to the construction of Regular Polygons with the aid of the sector, and of coefficients to facilitate their construction without it; also, of coefficients to aid in finding the area of the figure, the side only being given.

Names.	Number of sides.	Angle at center.	Angle at circum.	Perp'n side being 1.	Length of side rad. being 1.	Rad. of cir. side be- ing 1.	Rad. of cir. per. be- ing 1.	Area side being 1.
Triangle	3	120	60	.28868	1.782	.5773	2.	.433012
Square	4	90	90	.5	1.414	.7071	1.414	1.
Pentagon	5	72	108	.6882	1.175	.8506	1.238	1.720477
Hexagon	6	60	120	·.866	1.	1.	1.156	2.598076
Heptagon	7	51 3-7	128 4-7	1.0382	.8672	1.152	1.11	3.633912
Octagon	8	45	135	1.2071	.7654	1.3065	1.08	4.828427
Nonagon	9	40	140	1.3737	.684	1.4619	1.06	6.181824
Decagon	10	36	144	1.5388	.618	1.618	1.05	7.694208
Undecagon	11	32 8-11	147 3-11	1.7028	.5634	1.7747	1.04	9.36564
Dodecagon	12	30	150	1.866	.5176	1.9318	1.037	11.196152

NOTE.—" Angle at center " means the angle of radii passing from the center to the circumference or corners of the figure. "Angle at circumference" means the angle which any two adjoining sides make with each other.

THE CIRCLE AND ITS SECTIONS.

OBSERVATIONS AND DEFINITIONS.

1. The circle contains a greater area than any other plane figure bounded by the same perimeter or outline.

2. The areas of circles are to each other as the squares of their diameters; any circle twice the diameter of another contains four times the area of the other.

3. The radius of a circle is a straight line drawn from the center to the circumference.

4. The diameter of a circle is a straight line drawn

through the center and terminating both ways in the circumference.

5. A chord is a straight line joining any two points of the circumference.

6. The versed sine is a straight line joining the chord and the circumference.

7. An arc is any part of the circumference.

8. A semicircle is half the circle cut off by a diameter.

9. A segment is any portion of a circle cut off by a chord.

10. A sector is a part of a circle cut off by two radii.

General Rules in Relation to the Circle.

1. Multiply the diameter by 3.1416, the product is the circumference.

2. Multiply the circumference by .31831, the product is the diameter.

3. Multiply the square of the diameter by .7854 and the product is the area.

4. Multiply the square root of the area by 1.12837, the product is the diameter.

5. Multiply the diameter by .8862, the product is the side of a square of equal area.

6. Multiply the side of a square by 1.128, the product is the diameter of a circle of equal area.

Application of the Rules to Practical Purposes.

1. The diameter of a circle being 5 feet 6 inches, required its circumference.

5.5 × 3.1416 = 17.27880 feet, the circumference.

2. A straight line or the circumference of a circle being 17.27880 feet, required the circle's diameter corresponding thereto.

17.27880 $\times$ .31831 = 5.5000148280 feet, diameter.

3. The diameter of a circle is 9⅜ inches; what is its area in square inches?

$9.375^2 = 87.89$, etc., $\times$ .7854 = 69.029, etc., inches, the area.

4. What must the diameter of a circle be to contain an area equal to 69.029296875 square inches?

$\sqrt{69.02929}$, etc., $= 8.3091 \times 1.12837 = 9.375$, etc., or 9⅜ inches, the diameter.

5. The diameter of a circle is 15½ inches; what must each side of a square be to be equal in area to the given circle?

15.5 $\times$ .8862 = 13.73, etc., inches, length of side.

6. Each side of a square is 13.736 inches in length; what must the diameter of a circle be to contain an area equal to the given square?

13.736 $\times$ 1.128 = 15.49, etc., or 15½ inches, the diameter.

Any Chord and Versed Sine of a Circle being Given, to Find the Diameter.

RULE.—*Divide the sum of the squares of the versed size and one-half the chord by the versed sine; the quotient is the diameter of corresponding circle.*

7. The chord of a circle equals 8 feet and the versed sine equals 1½; required the circle's diameter.

$8^2 + 1.5^2 = 66.25 \div 1.5 = 44.16$ feet, the diameter.

8. In the curve of a railway I stretched a line 80 feet in length and the distance from the line to the curve I found to be 9 inches; required the circle's diameter.

$80^2 + .75^2 = 640.5625 \div 2 = 320.28$, etc., feet, the diameter.

To Find the Length of Any Arc of a Circle.

RULE.—*From eight times the chord of half the arc subtract the chord of the whole arc, and one-third of the remainder will be the length, nearly.*

Required the length of an arc, the chord of half the arc being 8½ feet and chord of whole arc 16 feet 8 inches.

$$8.5 \times 8 = 68.0 - 16.666 = \frac{51.334}{3} = 17.111^1/_3 \quad \text{cubic}$$

feet, the length of the arc.

To Find the Area of the Sector of a Circle.

RULE.—*Multiply the length of the arc by half the length of the radius.*

The length of the arc equals 9½ inches and the radii equal each 7 inches; required the area.

$$9.5 \times 3.5 = 33.25 \text{ inches, the area.}$$

To Find the Area of a Segment of a Circle.

RULE.—*Find the area of a sector whose arc is equal to that of the given segment, and if it be less than a semicircle subtract the area of the triangle formed by the chord of segment and radii of its extremities; but if more than a semicircle add area of triangle to the area of the sector, and the remainder or sum is the area of the segment.*

To Find the Area of the Space Contained Between Two Concentric Circles or the Area of a Circular Ring.

RULE I.—*Mutlply the sum of the inside and outside diameters by their difference and by .7854; the product is the area.*

RULE 2.—*The difference of the area of the two circles will be the area of the ring or space.*

Suppose the external circle equal 4 feet and the internal circle 2½ feet, required the area of space contained between them or area of a ring.

$4 + 2.5 = 6.5$ and $4 - 2.5 = 1.5$, hence, $6.5 \times 1.5 \times .7854 = 7.65$ feet, the area; or,

The area of 4 feet is 12.566; the area of 2.5 is 4.9081. (See table of areas of circles.)

$12.566 - 4.9081 = 7.6579$, the area.

To Find the Area of an Ellipse or Oval.

RULE.—*Multiply the diameters togther and their product by .7854.*

An oval is 20 x 15 inches, what are its superficial contents?

$20 \times 15 \times .7854 = 235.62$ inches, the area.

To Find the Circumference of an Ellipse or Oval.

RULE.—*Multiply half the sum of the two diameters by 3.1416 and the product will be the circumference.*

EXAMPLE.—An oval is 20 x 15 inches, what is the circumference.

$\dfrac{20 + 15}{2} = 17.5 \times 3.1416 = 54.978$ inches, the circumference.

OF CYLINDERS.

To Find the Convex Surface of a Cylinder.

RULE.—*Multiply the circumference by the hight or length, the product will be the surface.*

EXAMPLE.—The circumference of a cylinder is 6 feet

4 inches and its length 15 feet, required the convex surface.

$6.333 \times 15 = 94.995$ square feet, the surface.

OF CONES AND PYRAMIDS.

To Find the Convex Surface of a Right Cone or Pyramid.

RULE.—*Multiply the circumference of the base by the slant hight and half the product is the slant surface; if the surface of the entire figure is required, add the area of the base to the convex surface.*

EXAMPLE.—The base of a cone is 5 feet diameter and the slant hight is 7 feet, what is the convex surface?
$5 \times 3.1416 = 15.70$ circumference of the base and
$$\frac{15.70 \times 7}{2} = 54.95 \text{ square feet, the convex surface.}$$

To Find the Convex Surface of a Frustum of a Cone or Pyramid.

RULE.—*Multiply the sum of the circumference of the two ends by the slant hight and half the product will be the slant surface.*

The diameter of the top of the frustum of a cone is 3 feet, the base 5 feet, the slant hight 7 feet 3 inches; required the slant surface.

$$9.42 + 15.7 = \frac{25.12 \times 7.25}{2} = 91.06 \text{ square feet, slant surface.}$$

OF SPHERES.

To Find the Convex Surface of a Sphere or Globe.

RULE.—*Multiply the diameter of the sphere by its circumference and the product is its surface; or,*

Multiply the square of the diameter by 3.1416; the product is the surface.

What is the convex surface of a globe 6½ feet in diameter?

6.5 × 3.1416 × 6.5 = 132.73 square feet; or, 6.5^2 = 42.25 × 3.1416 = 132.73 square feet, the convex surface.

MENSURATION OF SOLIDS AND CAPACITIES OF BODIES.

To Find the Solidity or Capacity of Any Figures in the Cubical Form.

RULE.—*Multiply the length of any one side by its breadth and by the depth or distance to its opposite side, and the product is the solidity in equal terms of measurement.*

EXAMPLE.—The side of a cube is 20 inches; what is its solidity?

20 × 20 × 20 = 8000 cubic inches, or 4.6296 cubic feet, nearly.

A rectangular tank is in length 6 feet, in breadth 4½ feet and its depth 3 feet; required its capacity in cubic feet; also its capacity in United States standard gallons.

6 × 4.5 × 3 = 81 cubic feet; 81 × 1728 = 139,968 ÷ 231 = 605.92 gallons.

OF CYLINDERS.

To Find the Solidity of Cylinders.

RULE.—*Multiply the area of the base by the hight and the product is its solidity.*

EXAMPLE.—The base of a cylinder is 18 inches and hight 40 inches;

$$18^2 \times .7854 \times 40 = 10,178.7840 \text{ cubic inches.}$$

To Find the Contents in Gallons of Cylindrical Vessels.

RULE.—*Take the dimensions in inches and decimal parts of an inch. Square the diameter, multiply it by the hight, then multiply the product by .0034 for wine gallons, or by .002785 for beer gallons.*

EXAMPLE.—How many United States gallons will a cylinder contain whose diameter is 18 inches and length 30 inches?

$$18^2 \times 30 = 9720 \times .0034 = 33.04, \text{ etc., gallons.}$$

OF CONES AND PYRAMIDS.

To Find the Solidity of a Cone or a Pyramid.

RULE.—*Multiply the area of the base by the perpendicular hight and one-third the product will be the solidity.*

EXAMPLE.—The base of a cone is 2¼ feet and the hight is 3¾ feet, what is the solidity?

$$\frac{2.25^2 \times .7854 \times 3.75}{3} = 4.97 \text{ cubic feet, the solidity.}$$

To Find the Solidity of the Frustum of a Cone.

RULE.—*To the product of the diameters of the ends add one-third the square of the difference of the diameters; multiply the sum by .7854 and the product will be the mean area between the ends, which multiplied by the perpendicular hight of frustum gives the solidity.*

EXAMPLE.—The diameter of the large end of a frustum of a cone is 10 feet, that of the smaller end is 6 feet and the perpendicular hight 12 feet, what is its solidity?

$10 - 6 = 4^2 = 16 \div 3 = 5.333$ square of difference of ends; and $10 \times 6 + 5.333 = 65.333 \times .7854 \times 12 = 615.75$ cubic feet, the solidity.

To Find the Contents in U. S. Standard Gallons of the Frustum of a Cone.

RULE.—*To the product of the diameters, in inches and decimal parts of an inch, of the ends, add one-third the square of the difference of the diameters. Multiply the sum by the perpendicular hight in inches and decimal parts of an inch and multiply that product by .0034 for wine gallons, and by .002785 for beer gallons.*

EXAMPLE.—The diameter of the large end of a frustum of a cone is 8 feet, that of the smaller end is 4 feet and the perpendicular hight 10 feet; what are the contents in United States standard gallons?

$96 - 48 = 48^2 = 2304 \div 3 = 768$; $96 \times 48 + 768 = 5376 \times 120 \times .0034 = 2193.4$ gallons.

To Find the Solidity of the Frustum of a Pyramid.

RULE.—*Add to the areas of the two ends of the frustum the square root of their product, and this sum multi-*

plied by one-third of the perpendicular hight will give the solidity.

EXAMPLE.—What is the solidity of a hexagonal pyramid, a side of the large end being 12 feet, one of the smaller ends 6 feet and the perpendicular hight 8 feet?

374.122 × 93.53 = ^{V}34,991.63 = 187.06. 374.122 + 93.53 + 187.06 = $\dfrac{654.712 \times 8}{3}$ = 1745.898 cubic feet, solidity.

To Find the Solidity of a Sphere.

RULE.—*Multiply the cube of the diameter by .5236 and the product is the solidity.*

EXAMPLE.—What is the solidity of a sphere, the diameter being 20 inches?

20^3 = 8000 × .5236 = 4188.8 cubic inches, the solidity.

TABLES, RULES AND RECIPES.

BLACK SHEET IRON.

Black Sheets are rolled to the following Standard Gauges adopted by the United States, taking effect July 1, 1893.

Number of gauge.	THICKNESS. Approximate thickness in fractions of an inch.	Approximate thickness in decimal parts of an inch.	WEIGHT. Weight per square foot in ounces avoirdupois.	Weight per square foot in pounds avoirdupois.
10	9-64	.140625	90	5.625
11	1-8	.125	80	5.
12	7-64	.109375	70	4.375
13	3-32	.09375	60	3.75
14	5-64	.078125	50	3.125
15	9-128	.0703125	45	2.8125
16	1-16	.0625	40	2.5
17	9-160	.05625	36	2.25
18	1-20	.05	32	2.
19	7-160	.04375	28	1.75
20	3-80	.0375	24	1.50
21	11-320	.034375	22	1.375
22	1-32	.03125	20	1.25
23	9-320	.028125	18	1.125
24	1-40	.025	16	1.
25	7-320	.021875	14	.875
26	3-160	.01875	12	.75
27	11-640	.0171875	11	.6875
28	1-64	.015625	10	.625
29	9-640	.0140625	9	.5625
30	1-80	.0125	8	.5
31	7-640	.0109375	7	.4375
32	13-1280	.01015625	6½	.40625

A variation of 2½ per cent. either way is allowed.

PLATE IRON.

The following table gives the weight per square foot for iron plates 1-16 inch up to ½ inch thick.

Thickness.	Weight in lbs.	Thickness.	Weight in lbs.
1-16	2.50	5-16	12.50
1-3	5.00	3-8	15.00
3-16	7.50	7-16	17.50
1-4	10.00	1-2	20.00

Tables, Rules and Recipes.

WEIGHT OF SHEET LEAD.

The thickness of lead is in common determined or understood by the weight, the unit being that of a square or superficial foot; thus:

4 lbs. lead is 1-16 inch in thickness; 6 do. 1-10 do.; 7½ do. 1-8 do.; 11 do. 3-16 do.; 15 do. 1-4 do.

DECIMALS EQUIVALENT TO THE FRACTIONAL PARTS OF A POUND.

.03125	½ oz.	.28125	4½ oz.	.53125	8½ oz.	.78125	12½ oz.
.0625	1 "	.3125	5 "	.5625	9 "	.8125	13 "
.09375	1½ "	.34375	5½ "	.59375	9½ "	.84375	13½ "
.125	2 "	.375	6 "	.625	10 "	.875	14 "
.15625	2½ "	.40625	6½ "	.65625	10½ "	.90625	14½ "
.1875	3 "	.4375	7 "	.6875	11 "	.9375	15 "
.21875	3½ "	.46875	7½ "	.71875	11½ "	.96875	15½ "
.25	4 "	.5	8 "	.75	12 "	1.	16 "

DECIMALS EQUIVALENT TO THE FRACTIONAL PARTS OF AN INCH WHEN DIVIDED INTO 32 PARTS; LIKEWISE THE DECIMALS EQUIVALENT TO THE FRACTIONAL PARTS OF A FOOT.

Decimals.	Parts of an inch.		Decimals.	Parts of an inch.	Decimals.	Parts of a foot.
.03125	1-32		.53125	½ and 1-32	.01041	⅛
.0625	1-16		.5625	½ and 1-16	.02083	¼
.09375	3-32		.59375	½ and 3-32	.03125	⅜
.125	⅛		.625	⅝	.04166	½
.15625	⅛	and 1-32	.65625	⅝ and 1-32	.05208	⅝
.1875	⅛	and 1-16	.6875	⅝ and 1-16	.0625	¾
.21875	⅛	and 3-32	.71875	⅝ and 3-32	.07291	⅞
.25	¼		.75	¾	.0833	1
.28125	¼	and 1-32	.78125	¾ and 1-32	.1666	2
.3125	¼	and 1-16	.8125	¾ and 1-16	.25	3
.34375	¼	and 3-32	.84375	¾ and 3-32	.3333	4
.375	⅜		.875	⅞	.4166	5
.40625	⅜	and 1-32	.90625	⅞ and 1-32	.5	6
.4375	⅜	and 1-16	.9375	⅞ and 1-16	.5833	7
.46875	⅜	and 3-32	.96875	⅞ and 3-32	.6666	8
.5	½		1.	1 inch.	.75	9
					.8333	10
					.9166	11

TO ASCERTAIN THE WEIGHTS OF PIPES OF VARIOUS METALS, AND ANY DIAMETER REQUIRED.

Thick. Inch.	Wrought iron.	Copper.	Lead.	Thick. Inch.	Wrought iron.	Copper.	Lead.
1-32	.326	.38	.483	5-32	1.627	1.9	2.417
1-16	.653	.76	.967	3-16	1.95	2.28	2.9
3-32	.976	1.14	1.45	7-32	2.277	2.66	3.383
1-8	1.3	1.52	1.933	1-4	2.6	3.04	3.867

RULE.—*To the interior diameter of the pipe, in inches, add the thickness of the metal; multiply the sum by the decimal number opposite the required thickness and under the metal's name; also by the length of the pipe in feet; and the product is the weight of the pipe in pounds.*

1. Required the weight of a copper pipe whose interior diameter is 2½ inches, its length 20 feet, and the metal ⅛ inch in thickness.

$$2.25 + .125 = 2.375 \times 1.52 \times 20 = 72.2 \text{ pounds.}$$

WEIGHT OF GALVANIZED SHEETS.

	Ounces per square foot.		Ounces per square foot.		Ounces per square foot.
No. 14	52½	No. 20	26½	No. 26	14½
No. 15	47½	No. 21	24½	No. 27	13½
No. 16	42½	No. 22	22½	No. 28	12½
No. 17	38½	No. 23	20½	No. 29	11½
No. 18	34½	No. 24	18½	No. 30	10½
No. 19	30½	No. 25	16½		

ORDINARY DIMENSIONS OF GALVANIZED SHEETS.

Widths	40	38	36	34	32	30	28	26	24	22	20
Gauges.					Lengths.						
No. 14	..	..	96	96	96	96	96	96	96	96	96
Nos. 16 to 22	120	120	120	120	120	120	120	120	120	120	120
Nos. 23 and 24	96	96	96	96	108	120	120	120	120	108	108
Nos. 25 to 28	..	..	96	96	108	120	120	120	120	108	108
Nos. 29 and 30	..	..	..	..	..	..	96	96	96	96	..	..

WEIGHT PER FOOT OF LEAD PIPE.

Inside diameter. Ins.	AAA Brooklyn.		AA Ex. strong.		A Strong.		B Medium.		C Light.		D Ex. light.		E Fountain.	
	Lb.	Oz.	Lb.	Oz.	Lb.	Oz.	Lb.	Oz.	Lb.	Oz.	Lb.	Oz.	Lb.	Oz.
⅜	1	12	1	8	1	4	1	0	0	12	0	10	1	7
7-16	.	..	.	..	.	..	1	0	0	13	.	..	.	..
½	3	0	2	0	1	12	1	4	1	0	0	12	0	9
⅝	3	8	2	12	2	8	2	0	1	8	1	0	0	12
¾	4	12	3	8	3	0	2	4	1	12	1	4	1	0
1	6	0	4	12	4	0	3	4	2	8	2	0	1	8
1¼	6	12	5	12	4	12	3	12	3	0	2	8	2	0
1½	8	8	7	8	6	8	5	0	4	4	3	8	3	0
1¾	10	0	8	8	7	0	6	0	5	0	4	0	0	0
2	11	12	9	0	8	0	7	0	6	0	4	12	.	..

NET WEIGHT PER BOX TIN PLATES.

Basis 10 x 14, 225 sheets; or, 14 x 20, 112 sheets.

Size of sheets	Sheets per box	80 lb.	85 lb.	90 lb.	95 lb.	100 lb.	IC	IXL	IX	IXX	IXXX	IXXXX
Trade term		80 lb.	85 lb.	90 lb.	95 lb.	100 lb.	IC	IXL	IX	IXX	IXXX	IXXXX
Approximate wire gauge		No. 34	No. 33	No. 32	No. 31½	No. 31	No. 30	No. 28½	No. 28	No. 27	No. 26	No. 25
Weight per box, pounds		80	85	90	95	100	107	128	135	156	176	196
10 x 14	225	80	85	90	95	100	107	128	135	156	176	196
14 x 20	112	80	85	90	95	100	107	128	135	156	176	196
20 x 28	112	160	170	180	190	200	214	256	270	312	352	392
10 x 20	225	114	121	129	136	143	153	183	193	223	251	280
11 x 11	225	69	73	78	82	86	92	111	117	135	152	169
11 x 22	225	138	147	156	164	172	184	222	234	270	304	339
11½ x 23	225	151	161	170	179	189	202	242	255	295	333	370
12 x 12	225	82	87	93	98	103	110	132	139	160	181	202
12 x 24	112	82	87	93	98	103	110	132	139	160	181	202
13 x 13	225	97	103	109	115	121	129	154	163	188	212	236
13 x 26	112	97	103	109	115	121	129	154	163	188	212	236
14 x 14	225	112	119	126	133	140	150	179	189	218	246	274
14 x 28	112	112	119	126	133	140	150	179	189	218	246	274
15 x 15	225	129	137	145	153	161	172	206	217	251	283	315
16 x 16	225	146	155	165	174	183	196	234	247	285	322	358
17 x 17	225	165	175	186	196	206	221	264	279	322	363	405
18 x 18	112	93	98	104	110	116	124	148	156	180	204	227
19 x 19	112	103	110	116	122	129	138	165	174	201	227	253
20 x 20	112	114	121	129	136	143	153	183	193	223	251	280
21 x 21	112	126	134	142	150	158	169	202	213	246	277	309
22 x 22	112	138	147	156	164	172	184	221	234	270	304	339

NET WEIGHT PER BOX TIN PLATES.

Basis 10 x 14, 225 sheets; or, 14 x 20, 112 sheets.

Trade term		80 lb.	85 lb.	90 lb.	95 lb.	100 lb.	IC	IXL	IX	IXX	IXXX	IXXXX
Approximate wire gauge		No. 34	No. 33	No. 32	No. 31½	No. 31	No. 30	No. 28½	No. 28	No. 27	No. 26	No. 25
Weight per box, pounds		80	85	90	95	100	107	128	135	156	176	196
Size of sheets.	Sheets per box.											
23 x 23	112	151	161	170	179	189	202	242	255	295	333	370
24 x 24	112	164	175	185	195	204	220	263	278	321	362	404
26 x 26	112	193	205	217	229	241	258	309	326	377	424	472
13½ x 19½	112	75	80	85	89	94	109	120	127	147	165	183
14 x 18¾	124	83	88	93	98	103	110	132	139	161	182	202
14 x 19¼	120	85	88	93	98	103	110	132	139	161	182	202
14 x 21	112	84	89	95	100	105	112	134	142	164	185	206
14 x 22	112	88	94	99	105	110	118	141	149	172	194	216
15 x 21	112	90	95	101	107	113	120	144	152	176	197	220
16 x 20	112	91	97	103	109	114	122	146	154	178	201	224
14 x 31	112	124	132	140	147	155	166	198	209	242	273	304

Approximate wire gauge D plates		No. 28		No. 25	No. 24	No. 23	No. 22
12½ x 17D	100	94	......	122	142	162	182
17 x 25D	50	94	......	122	142	162	182
15 x 21D	100	140	......	181	211	241	271

Taggers iron and tin.	10 x 14 Sheets per box.	10 x 14 Pounds per box.	14 x 20 Sheets per box.	14 x 20 Pounds per box.	20 x 28 Sheets per box.	20 x 28 Pounds per box.	20 x 40 Sheets per box.	20 x 40 Pounds per box.
No. 30 W G			112	107	112	214	79	224
No. 32 W G			128	112	128	224	79	180
No. 34 W G	300	112	150	112	150	224	79	160
No. 36 W G	360	112	180	112	180	224		
No. 38 W G	450	112	225	112	225	224		

SHEET ZINC.

APPROXIMATE WEIGHT PER SHEET.

Numbers	Weight per sq. foot	Approx. thickness in inches	Size of sheet	Sq.ft. per sht.

Number	Wt. per sq.ft.	Thick. (in)	24 x 84 (14.)	26 x 84 (15.2)	28 x 84 (16.3)	30 x 84 (17.5)	32 x 84 (18.7)	34 x 84 (19.9)	36 x 84 (21.)	36 x 96 (24.)	36 x 108 (27.)	40 x 84 (23.4)	40 x 96 (26.8)	44 x 84 (25.7)	46 x 90 (28.7)	48 x 84 (28.)	48 x 96 (32.)	50 x 108 (37.5)	52 x 84 (30.4)
4	.30	.008	4.2	4.6	4.9	5.3	5.6	6.0	6.3	7.2	8.1	7.	8.	7.7	8.6	8.4	9.6	11.3	9.1
5	.37	.010	5.2	5.6	6.	6.5	6.9	7.4	7.8	8.9	10.	8.7	9.9	9.5	10.6	10.4	11.9	13.9	11.3
6	.45	.012	6.3	6.9	7.4	7.9	8.4	9.	9.5	10.8	12.2	10.6	12.1	11.6	12.9	12.6	14.4	16.9	13.7
7	.52	.014	7.3	7.9	8.5	9.1	9.7	10.4	10.9	12.5	14.1	12.2	14.	13.4	14.9	14.6	16.7	19.5	15.8
8	.60	.016	8.4	9.1	9.8	10.5	11.2	12.	12.6	14.4	16.2	14.1	16.1	15.4	17.2	16.8	19.2	22.5	18.3
9	.67	.018	9.4	10.2	10.9	11.8	12.6	13.4	14.1	16.1	18.1	15.7	18.	17.2	19.2	18.8	21.5	25.1	20.4
10	.75	.020	10.5	11.4	12.2	13.2	14.1	15.	15.3	18.	20.3	17.6	20.1	19.3	21.5	21.	24.	28.2	22.8
11	.90	.024	12.6	13.7	14.7	15.8	16.9	18.	18.9	21.6	24.3	21.	24.1	23.1	25.8	25.2	28.8	33.8	27.4
12	1.05	.028	14.7	16.	17.1	18.4	19.7	20.9	22.	25.2	28.4	24.6	28.1	27.	30.1	29.4	33.6	39.3	31.9
13	1.20	.032	16.8	18.3	19.6	21.	22.5	23.9	25.2	28.8	32.4	28.1	32.2	30.8	34.4	33.6	38.4	45.	36.5
14	1.35	.036	18.9	20.5	22.	23.6	25.3	26.9	28.4	32.4	36.5	31.6	36.2	34.7	38.7	37.8	43.2	50.7	41.
15	1.50	.040	21.	22.8	24.5	26.2	28.	29.9	31.5	36.	40.5	35.1	40.2	38.6	43.	42.	48.	56.3	45.6
16	1.68	.045	23.5	25.6	27.4	29.4	31.4	33.4	35.3	40.3	45.4	39.3	45.	43.2	48.2	47.	53.8	63.	51.
17	1.87	.050	26.2	28.4	30.5	32.8	35.	37.1	39.3	44.9	50.5	43.8	50.1	48.1	53.7	52.4	59.9	70.1	56.9
18	2.06	.055	28.9	31.3	33.6	36.1	38.5	41.	43.2	49.5	55.6	48.2	55.2	53.	59.1	57.7	65.9	77.3	62.6
19	2.25	.060	31.5	34.2	36.7	39.4	42.	44.8	47.2	54.	60.7	52.6	60.3	57.8	64.6	63.	72.	84.4	68.4
20	2.62	.070	36.7	39.9	42.7	45.8	49.	52.2	55.	62.8	70.7	61.3	70.2	67.4	75.2	73.4	83.9	98.3	79.6
21	3.00	.080	42.	45.6	48.9	52.5	56.1	59.7	63.	72.	81.	70.2	80.4	77.1	86.1	84.	96.	112.5	91.2
22	3.37	.090	47.2	51.2	54.9	59.	63.	67.	70.8	80.9	91.	78.8	90.3	86.6	96.7	94.4	107.8	126.4	102.5

Casks average about 600 pounds each. No. 4 to No. 17. Boxes average about 500 pounds. No. 18 and heavier.

RELATIVE WEIGHTS OF ALUMINUM AND COPPER SHEETS.

ROLLED ALUMINUM has a specific gravity of 2.72. One cubic foot weighs $169\frac{510}{1000}$ lbs. One square foot of one inch thick weighs $14\frac{126}{1000}$ lbs. Rolled Copper is 3.283 times heavier than similar sections of Rolled Aluminum.

Stub's gauge (nearest) No.	Thickness in decimal parts of 1 inch.	Oz. per square foot of copper.	Oz. per square foot of aluminum of same thickness.	Sheets 14 x 48 weight in pounds of copper.	Sheets 14 x 48 weight in pounds of aluminum of same thickness.	Sheets 24 x 48 weight in pounds of copper.	Sheets 24 x 48 weight in pounds of aluminum of same thickness.	Sheets 30 x 60 weight in pounds of copper.	Sheets 30 x 60 weight in pounds of aluminum of same thickness.	Sheets 36 x 72 weight in pounds of copper.	Sheets 36 x 72 weight in pounds of aluminum of same thickness.	Sheets 48 x 72 weight in pounds of copper.	Sheets 48 x 72 weight in pounds of aluminum of same thickness.
35	.00537	4	1.22	1.16	0.35	2	0.61	3.12	0.96	4.50	1.38	6	1.83
33	.00806	6	1.83	1.75	0.53	3	0.92	4.68	1.43	6.75	2.06	9	2.75
31	.0107	8	2.44	2.33	0.71	4	1.22	6.25	1.91	9	2.75	12	3.66
29	.0134	10	3.05	2.91	0.89	5	1.53	7.81	2.38	11.25	3.43	15	4.57
27	.0161	12	3.66	3.50	1.07	6	1.83	9.37	2.86	13.50	4.12	18	5.49
26	.0183	14	4.27	4.08	1.25	7	2.14	10.93	3.33	15.75	4.80	21	6.40
24	.0215	16	4.88	4.66	1.42	8	2.44	12.50	3.81	18	5.49	24	7.32
23	.0242	18	5.49	5.25	1.60	9	2.75	14.06	4.29	20.25	6.17	27	8.23
22	.0269	20	6.10	5.83	1.78	10	3.05	15.62	4.76	22.50	6.86	30	9.14
21	.0322	24	7.32	7	2.14	12	3.66	18.75	5.72	27	8.23	36	11.00
19	.0430	32	9.75	9.33	2.85	16	4.88	25	7.62	36	11.00	48	14.70
18	.0538	40	12.20	11.66	3.56	20	6.10	31.25	9.52	45	13.75	60	18.30
16	.0645	48	14.65	14	4.27	24	7.32	37.50	11.45	54	16.50	72	22.00
15	.0754	56	17.10	16.33	4.98	28	8.53	43.75	13.35	63	19.20	84	25.60
14	.0860	64	19.50	18.66	5.69	32	9.75	50	15.30	72	21.95	96	29.30
13	.095	70	21.35	...	..	35	10.70	55	16.80	79	24.10	105	32.00
12	.109	81	24.70	...	..	40½	12.40	63	19.20	91	27.75	122	37.20
11	.120	89	27.15	...	..	44½	13.60	70	21.35	100	30.50	134	40.85
10	.134	100	30.50	...	..	50	15.30	78	23.80	112	34.20	150	45.70
9	.148	110	33.55	...	..	55	16.80	86	26.20	124	37.80	165	50.30
8	.165	123	37.50	...	..	61	18.60	96	29.30	138	42.10	184	56.10
7	.180	134	40.85	...	..	67	20.40	105	32.00	151	46.00	201	61 30
6	.203	151	46.00	...	..	75½	23.00	118	36.00	170	51.80	227	69.20
5	.220	164	50.00	...	..	82	25.00	128	39.00	184	56.10	246	75.00
4	.238	177	53.95	...	..	88½	27.00	138	42.10	199	60.70	266	81.10
3	.259	193	64.30	...	..	96	29.30	151	46.00	217	66.10	289	88.10
2	.284	211	67.95	...	..	105½	32.20	165	50.30	238	72.50	317	96.60
1	.300	223	77.10	...	..	111½	34.00	174	53.10	251	76.50	335	102.20
0	.340	253	...	...	..	126½	38.60	198	60.40	285	86.90	380	116.00

One ounce per square foot aluminum sheet is 0.0044 inch thick and corresponds to about No. 37 B. & S. gauge.

SHEET COPPER.

Official table adopted by the Association of Copper Manufacturers of the United States.

Rolled copper has specific gravity of 8.93. One cubic foot weighs $558^{125}/_{1000}$ pounds. One square foot, of 1 inch thick, weighs $46^{51}/_{100}$ pounds.

Stubs' gauge (nearest) number.	Thickness in decimal parts of 1 inch.	Ounces per square foot.	Sheets 14 x 48, weight in lbs.	Sheets 24 x 48, weight in lbs.	Sheets 30 x 60, weight in lbs.	Sheets 36 x 72, weight in lbs.	Sheets 48 x 72, weight in lbs.
35	.00537	4	1.16	2	3.12	4.50	6
33	.00806	6	1.75	3	4.68	6.75	9
31	.0107	8	2.33	4	6.25	9	12
29	.0134	10	2.91	5	7.81	11.25	15
27	.0161	12	3.50	6	9.37	13.50	18
26	.0188	14	4.08	7	10.93	15.75	21
24	.0215	16	4.66	8	12.50	18	24
23	.0242	18	5.25	9	14.06	20.25	27
22	.0269	20	5.83	10	15.62	22.50	30
21	.0322	24	7	12	18.75	27	36
19	.0430	32	9.33	16	25	36	48
18	.0538	40	11.66	20	31.25	45	60
16	.0645	48	14	24	37.50	54	72
15	.0754	56	16.33	28	43.75	63	84
14	.0860	64	18.66	32	50	72	96
13	.095	70	....	35	55	79	105
12	.109	81	....	40½	63	91	122
11	.120	89	....	44½	70	100	134
10	.134	100	....	50	78	112	150
9	.148	110	....	55	86	124	165
8	.165	123	....	61	96	138	184
7	.180	134	....	67	105	151	201
6	.203	151	....	75½	118	170	227
5	.220	164	....	82	128	184	246
4	.238	177	....	88½	138	199	266
3	.259	193	....	96	151	217	289
2	.284	211	....	105½	165	238	317
1	.300	223	....	111½	174	251	335
0	.340	253	....	126½	198	285	380

TABLES

OF THE

CIRCUMFERENCES OF CIRCLES,

TO THE

NEAREST FRACTION OF PRACTICAL MEASUREMENT;

ALSO,

THE AREAS OF CIRCLES, IN INCHES AND DECIMAL PARTS, LIKEWISE IN FEET AND DECIMAL PARTS, AS MAY BE REQUIRED.

Rules that may render the following tables more generally useful.

1. Any of the areas in inches, multiplied by .052, or the areas in feet multiplied by 7.48, the product is the number of gallons at 1 foot in depth.

2. Any of the areas in feet, multiplied by .03704, the product equals the number of cubic yards at 1 foot in depth.

Dia. in inch.	Circum. in inch.	Area in sq. inch.	Side of = sq.	Dia. in inch.	Cir. in ft. in.	Area in sq. inch.	Area in sq. ft.
1-16	.196	.0030	.0554	1 in.	3⅛	.7854	⅞
1-8	.392	.0122	.1107	1⅛	3½	.9940	⅞ and 3-32
3-16	.589	.0276	.1661	1¼	3⅞	1.227	1 in.
1-4	.785	.0490	.2115	1⅜	4¼	1.484	1 3-16
5-16	.981	.0767	.2669	1½	4⅝	1.767	1 5-16
3-8	1.178	.1104	.3223	1⅝	5⅛	2.074	1 7-16
7-16	1.374	.1503	.3771	1¾	5½	2.405	1 9-16
				1⅞	5⅞	2.761	1 11-16
1-2	1.570	.1963	.4331	2 in.	6¼	3.141	1¾
9-16	1.767	.2485	.4995	2⅛	6⅝	3.546	1⅞
5-8	1.963	.3068	.5438	2¼	7	3.976	2 in.
11-16	2.159	.3712	.6093	2⅜	7⅜	4.430	2⅛
3-4	2.356	.4417	.6646	2½	7¾	4.908	2 3-16
13-16	2.552	.5185	.7200	2⅝	8¼	5.412	2 5-16
7-8	2.748	.6013	.7754	2¾	8⅝	5.939	2 7-16
15-16	2.945	.6903	.8308	2⅞	9	6.491	2 9-16

Dia. in. inch.	Cir. in inch	Area in sq. inch.	Side of = sq.
3 in	9⅜	7.068	2⅝
3⅛	9¾	7.669	2¾
3¼	10¼	8.295	2⅞
3⅜	10⅝	8.946	3 in.
3½	11	9.621	3⅛
3⅝	11⅜	10.320	3¼
3¾	11¾	11.044	3⅜
3⅞	12⅛	11.793	3 7-16

Dia. in inch.	Cir. in ft.	in.	Area in sq. inch.	Area in sq. ft.
4 in.	1	0½	12.566	.0879
4⅛	1	0⅞	13.364	.0935
4¼	1	1⅜	14.186	.0993
4⅜	1	1¾	15.033	.1052
4½	1	2⅛	15.904	.1113
4⅝	1	2½	16.800	.1176
4¾	1	2⅞	17.720	.1240
4⅞	1	3¼	18.665	.1306
5 in.	1	3⅝	19.635	.1374
5⅛	1	4⅛	20.629	.1444
5¼	1	4½	21.647	.1515
5⅜	1	4⅞	22.690	.1588
5½	1	5¼	23.758	.1663
5⅝	1	5⅝	24.850	.1739
5¾	1	6	25.967	.1817
5⅞	1	6⅜	27.108	.1897
6 in.	1	6¾	28.274	.1979
6⅛	1	7¼	29.464	.2062
6¼	1	7⅝	30.679	.2147
6⅜	1	8	31.919	.2234
6½	1	8⅜	33.183	.2322
6⅝	1	8¾	34.471	.2412
6¾	1	9⅛	35.784	.2504
6⅞	1	9½	37.122	.2598
7 in.	1	10	38.484	.2693
7⅛	1	10⅜	39.871	.2791
7¼	1	10¾	41.282	.2889
7⅜	1	11⅛	42.718	.2990
7½	1	11½	44.178	.3092
7⅝	1	11⅞	45.663	.3196
7¾	2	0⅜	47.173	.3299
7⅞	2	0¾	47.707	.3409
8 in	2	1⅛	50.265	.3518
8⅛	2	1½	51.848	.3629
8¼	2	1⅞	53.456	.3741
8⅜	2	2¼	55.088	.3856
8½	2	2⅝	56.745	.3972
8⅝	2	3	58.426	.4089
8¾	2	3⅜	60.132	.4209
8⅞	2	3⅞	61.862	.4330
9 in.	2	4¼	63.617	.4453
9⅛	2	4⅝	65.396	.4517
9¼	2	5	67.200	.4704
9⅜	2	5⅜	69.029	.4832
9½	2	5¾	70.882	.4961
9⅝	2	6¼	72.759	.5093
9¾	2	6⅝	74.662	.5226
9⅞	2	7	76.588	.5361

Dia. in inch.	Cir. in ft.	in.	Area in sq. inch.	Area in sq. ft.
10 in.	2	7⅜	78.540	.5497
10⅛	2	7¾	80.515	.5636
10¼	2	8⅛	82.516	.5776
10⅜	2	8½	84.540	.5917
10½	2	8⅞	86.590	.6061
10⅝	2	9⅜	88.664	.6206
10¾	2	9¾	90.762	.6353
10⅞	2	10⅛	92.855	.6499
11 in.	2	10½	95.033	.6652
11⅛	2	10⅞	97.205	.6874
11¼	2	11¼	99.402	.6958
11⅜	2	11¾	101.623	.7143
11½	3	0⅛	103.869	.7290
11⅝	3	0½	106.139	.7429
11¾	3	0⅞	108.434	.7590
11⅞	3	1¼	110.753	.7752
12 in.	3	1⅝	113.097	.7916
12⅛	3	2	115.466	.8082
12¼	3	2½	117.859	.8250
12⅜	3	2⅞	120.276	.8419
12½	3	3¼	122.718	.8590
12⅝	3	3¾	125.185	.8762
12¾	3	4	127.676	.8937
12⅞	3	4⅜	130.192	.9113
13 in.	3	4¾	132.732	.9291
13⅛	3	5¼	135.297	.9470
13¼	3	5⅝	137.886	.9642
13⅜	3	6	140.500	.9835
13½	3	6⅜	143.139	1.0019
13⅝	3	6¾	145.802	1.0206
13¾	3	7⅛	148.489	1.0294
13⅞	3	7½	151.201	1.0584
14 in.	3	7⅞	153.938	1.0775
14⅛	3	8⅜	156.699	1.0968
14¼	3	8¾	159.485	1.1193
14⅜	3	9⅛	162.295	1.1360
14½	3	9½	165.130	1.1569
14⅝	3	9⅞	167.989	1.1749
14¾	3	10¼	170.873	1.1961
14⅞	3	10¾	173.782	1.2164
15 in.	3	11⅛	176.715	1.2370
15⅛	3	11½	179.672	1.2577
15¼	3	11⅞	182.654	1.2785
15⅜	4	0¼	185.661	1.2996
15½	4	0⅝	188.692	1.3208
15⅝	4	1	191.748	1.3422
15¾	4	1½	194.828	1.3637
15⅞	4	1⅞	197.933	1.3855
16 in.	4	2¼	201.062	1.4074
16⅛	4	2⅝	204.216	1.4295
16¼	4	3	207.394	1.4517
16⅜	4	3⅜	210.597	1.4741
16½	4	3¾	213.825	1.4967
16⅝	4	4¼	217.077	1.5195
16¾	4	4⅝	220.353	1.5424
16⅞	4	5	223.654	1.5655

Dia. in inch.	Cir. in ft.	in.	Area in sq. inch.	Area in sq. ft.	Dia. in ft.	in.	Cir. in ft.	in.	Area in sq. inch.	Area in sq. ft.
17 in.	4	5⅜	226.980	1.5888	2	0	6	3⅜	452.290	3.1418
17⅛	4	5¾	230.330	1.6123	2	0¼	6	4⅛	461.864	3.2075
17¼	4	6⅛	233.705	1.6359	2	0½	6	4⅞	471.436	3.2731
17⅜	4	6½	237.104	1.6597	2	0¾	6	5¾	481.106	3.3410
17½	4	6⅞	240.528	1.6836	2	1	6	6½	490.875	3.4081
17⅝	4	7⅜	243.977	1.7078	2	1¼	6	7¼	500.741	3.4775
17¾	4	7¾	247.450	1.7321	2	1½	6	8⅛	510.706	3.5468
17⅞	4	8⅛	250.947	1.7566	2	1¾	6	8⅞	520.769	3.6101
18 in.	4	8½	254.469	1.7812	2	2	6	9⅝	530.930	3.6870
18⅛	4	8⅞	258.016	1.8061	2	2¼	6	10½	541.189	3.7583
18¼	4	9¼	261.587	1.8311	2	2½	6	11¼	551.547	3.8302
18⅜	4	9¾	265.182	1.8562	2	2¾	7	0	562.002	3.9042
18⅜	4	10⅛	268.803	1.8816	2	3	7	0¾	572.556	3.9761
18⅝	4	10½	272.447	1.9071	2	3¼	7	1⅝	583.208	4.0500
18¾	4	10⅞	276.117	1.9328	2	3½	7	2⅜	593.958	4.1241
18⅞	4	11¼	279.811	1.9586	2	3¾	7	3⅛	604.807	4.2000
19 in.	4	11⅝	283.529	1.9847	2	4	7	3⅞	615.753	4.2760
19⅛	5	0	287.272	1.9941	2	4¼	7	4¾	626.798	4.3521
19¼	5	0½	291.039	2.0371	2	4½	7	5½	637.941	4.4302
19⅜	5	0⅞	294.831	2.0637	2	4¾	7	6¼	649.182	4.5083
19½	5	1¼	298.648	2.0904	2	5	7	7	660.521	4.5861
19⅝	5	1⅝	302.489	2.1172	2	5¼	7	7⅛	671.958	4.6665
19¾	5	2	306.355	2.1443	2	5½	7	8⅝	683.494	4.7467
19⅞	5	2⅜	310.245	2.1716	2	5¾	7	9½	695.128	4.8274
20 in.	5	2⅞	314.160	2.1990	2	6	7	10¼	706.860	4.9081
20⅛	5	3¼	318.099	2.2265	2	6¼	7	11	718.690	4.9901
20¼	5	3⅝	322.063	2.2543	2	6½	7	11¾	730.618	5.0731
20⅜	5	4	326.051	2.2822	2	6¾	8	0⅝	742.644	5.1573
20½	5	4⅜	330.064	2.3103	2	7	8	1⅜	754.769	5.2278
20⅝	5	4¾	334.101	2.3386	2	7¼	8	2⅛	766.992	5.3264
20¾	5	5⅛	338.163	2.3670	2	7½	8	2⅞	779.313	5.4112
20⅞	5	5½	342.250	2.3956	2	7¾	8	3¾	791.732	5.4982
21 in.	5	5⅞	346.361	2.4244	2	8	8	4½	804.249	5.5850
21⅛	5	6⅜	350.497	2.4533	2	8¼	8	5⅜	816.865	5.6729
21¼	5	6¾	354.657	2.4824	2	8½	8	6⅛	829.578	5.7601
21⅜	5	7⅛	358.841	2.5117	2	8¾	8	6⅞	842.390	5.8491
21½	5	7½	363.051	2.5412	2	9	8	7⅝	855.300	5.9398
21⅝	5	7⅞	367.284	2.5708	2	9¼	8	8½	868.308	6.0291
21¾	5	8¼	371.543	2.6007	2	9½	8	9¼	881.415	6.1201
21⅞	5	8¾	375.826	2.6306	2	9¾	8	10	894.619	6.2129
22 in.	5	9⅛	380.133	2.6608	2	10	8	10¾	907.922	6.3051
22⅛	5	9½	384.465	2.6691	2	10¼	8	11½	921.323	6.3981
22¼	5	9⅞	388.822	2.7016	2	10½	9	0⅜	934.822	6.4911
22⅜	5	10¼	393.203	2.7224	2	10¾	9	1⅛	948.419	6.5863
22½	5	10⅝	397.608	2.7632	2	11	9	1⅞	962.115	6.6815
22⅝	5	11	402.038	2.7980	2	11¼	9	2¾	975.908	6.7772
22¾	5	11½	406.493	2.8054	2	11½	9	3½	989.800	6.8738
22⅞	5	11⅞	410.972	2.8658	2	11¾	9	4¼	1003.79	6.9701
23 in.	6	0¼	415.476	2.8903	3	0	9	5	1017.87	7.0688
23⅛	6	0⅝	420.004	2.9100	3	0¼	9	5⅞	1032.06	7.1671
23¼	6	1	424.557	2.9518	3	0½	9	6⅝	1046.35	7.2664
23⅜	6	1⅜	429.135	2.9937	3	0¾	9	7½	1060.73	7.3662
23½	6	1¾	433.737	3.0129	3	1	9	8¼	1075.21	7.4661
23⅝	6	2¼	438.363	3.0261	3	1¼	9	9	1089.79	7.5671
23¾	6	2⅝	443.014	3.0722	3	1½	9	9⅞	1104.46	7.6691
23⅞	6	3	447.690	3.1081	3	1¾	9	10½	1119.24	7.7791

Dia. in ft.	in.	Cir. in ft.	in.	Area in sq. inch.	Area in sq. ft.
3	2	9	11⅜	1134.12	7.8681
3	2¼	10	0⅛	1149.09	7.9791
3	2½	10	0⅞	1164.16	8.0846
3	2¾	10	1¾	1179.32	8.1891
3	3	10	2½	1194.59	8.2951
3	3¼	10	3¼	1209.95	8.4026
3	3½	10	4	1225.42	8.5091
3	3¾	10	4⅞	1240.98	8.6171
3	4	10	5⅝	1256.64	8.7269
3	4¼	10	6⅜	1272.39	8.8361
3	4½	10	7¼	1288.25	8.9462
3	4¾	10	8	1304.20	9.0561
3	5	10	8¾	1320.25	9.1686
3	5¼	10	9½	1336.40	9.2112
3	5½	10	10⅜	1352.65	9.3936
3	5¾	10	11⅛	1369.00	9.5061
3	6	10	11⅞	1385.44	9.6212
3	6¼	11	0¾	1401.98	9.7364
3	6½	11	1½	1418.62	9.8518
3	6¾	11	2¼	1435.36	9.9671
3	7	11	3	1452.20	10.084
3	7¼	11	3⅞	1469.14	10.202
3	7½	11	4⅝	1486.17	10.320
3	7¾	11	5⅜	1503.30	10.439
3	8	11	6¼	1530.53	10.559
3	8¼	11	7	1537.86	10.679
3	8½	11	7¾	1555.28	10.800
3	8¾	11	8½	1572.81	10.922
3	9	11	9⅜	1590.43	11.044
3	9¼	11	10⅛	1608.15	11.167
3	9½	11	10⅞	1625.97	11.291
3	9¾	11	11¾	1643.89	11.415
3	10	12	0½	1661.90	11.534
3	10¼	12	1¼	1680.02	11.666
3	10½	12	2	1698.23	11.793
3	10¾	12	2⅝	1716.54	11.920
3	11	12	3⅛	1734.94	12.048
3	11¼	12	4⅝	1753.45	12.176
3	11½	12	5¼	1772.05	12.305
3	11¾	12	6	1790.76	12.435
4	0	12	6¾	1809.56	12.566
4	0¼	12	7½	1828.46	12.697
4	0½	12	8⅜	1847.45	12.829
4	0¾	12	9⅛	1866.55	12.962
4	1	12	9⅞	1885.74	13.095
4	1¼	12	10⅝	1905.03	13.229
4	1½	12	11½	1924.42	13.304
4	1¾	13	0¼	1943.91	13.499
4	2	13	1	1963.50	13.635
4	2¼	13	1⅞	1983.18	13.772
4	2½	13	2½	2002.96	13.909
4	2¾	13	3⅜	2022.84	14.047
4	3	13	4¼	2042.82	14.186
4	3¼	13	5	2062.90	14.325
4	3½	13	5¾	2083.07	14.465
4	3¾	13	6½	2103.35	14.606
4	4	13	7⅜	2123.72	14.748
4	4¼	13	8⅛	2144.19	14.890
4	4½	13	8⅞	2164.75	15.033
4	4¾	13	9¾	2185.42	15.176
4	5	13	10½	2206.18	15.320
4	5¼	13	11¼	2227.05	15.465
4	5½	14	0	2248.01	15.611
4	5¾	14	0⅞	2269.06	15.757
4	6	14	1⅝	2290.22	15.904
4	6¼	14	2⅜	2311.48	16.051
4	6½	14	3¼	2332.83	16.200
4	6¾	14	4	2354.28	16.349
4	7	14	4¾	2375.83	16.498
4	7¼	14	5½	2397.48	16.649
4	7½	14	6⅜	2419.22	16.800
4	7¾	14	7⅛	2441.07	16.951
4	8	14	7⅞	2463.01	17.104
4	8¼	14	8⅝	2485.05	17.256
4	8½	14	9½	2507.19	17.411
4	8¾	14	10¼	2529.42	17.565
4	9	14	11	2551.76	17.720
4	9¼	14	11⅞	2574.19	17.876
4	9½	15	0⅝	2596.72	18.033
4	9¾	15	1⅜	2619.35	18.189
4	10	15	2¼	2642.08	18.347
4	10¼	15	2⅞	2664.91	18.506
4	10½	15	3¾	2687.83	18.665
4	10¾	15	4½	2710.85	18.825
4	11	15	5¼	2733.97	18.965
4	11¼	15	6⅛	2757.19	19.147
4	11½	15	6⅞	2780.51	19.309
4	11¾	15	7¾	2803.92	19.471
5	0	15	8½	2827.44	19.635
5	0¼	15	9¼	2851.05	19.798
5	0½	15	10	2874.76	19.963
6	0¾	15	10¾	2898.56	20.128
5	1	15	11⅝	2922.47	20.294
5	1¼	16	0⅜	2946.47	20.461
5	1½	16	1¼	2970.57	20.629
5	1¾	16	1⅞	2994.77	20.797
5	2	16	2¾	3019.07	20.965
5	2¼	16	3½	3043.47	21.135
5	2½	16	4¼	3067.96	21.305
5	2¾	16	5⅛	3092.56	21.476
5	3	16	5⅞	3117.25	21.647
5	3¼	16	6¼	3142.04	21.819
5	3½	16	7½	3166.92	21.992
5	3¾	16	8¼	3191.91	22.166
5	4	16	9	3216.99	22.333
5	4¼	16	9¾	3242.17	22.515
5	4½	16	10⅝	3267.46	22.621
5	4¾	16	11⅜	3292.83	22.866
5	5	17	0⅛	3318.31	23.043
5	5¼	17	0⅞	3343.88	23.221
5	5½	17	1¾	3369.56	23.330
5	5¾	17	2½	3395.33	23.578

Dia. in ft.	in.	Cir. in ft.	in.	Area in sq. inch.	Area in sq. ft.
5	6	17	3⅝	3421.20	23.758
5	6¼	17	4⅛	3447.16	23.938
5	6½	17	4⅜	3473.23	24.119
5	6¾	17	5⅞	3499.39	24.301
5	7	17	6½	3525.26	24.483
5	7¼	17	7¼	3552.01	24.666
5	7½	17	8	3578.47	24.850
5	7¾	17	8¾	3605.03	25.034
5	8	17	9⅝	3631.68	25.220
5	8¼	17	10⅜	3658.44	25.405
5	8½	17	11⅛	3685.29	25.592
5	8¾	17	11⅞	3712.24	25.779
5	9	18	0¾	3739.28	25.964
5	9¼	18	1½	3766.43	26.155
5	9½	18	2¼	3793.67	26.344
5	9¾	18	3⅛	3821.02	26.534
5	10	18	3⅞	3848.46	26.725
5	10¼	18	4⅜	3875.99	26.916
5	10½	18	5½	3903.63	27.108
5	10¾	18	6¼	3931.36	27.301
5	11	18	7	3959.20	27.494
5	11¼	18	7¾	3987.13	27.688
5	11½	18	8⅝	4015.16	27.883
5	11¾	18	9⅜	4043.28	28.078
6	0	18	10⅛	4071.51	28.274
6	0¼	18	10⅞	4099.83	28.471
6	0½	18	11¾	4128.25	28.663
6	0¾	19	0½	4156.77	28.866
6	1	19	1¼	4185.39	29.064
6	1¼	19	2⅛	4214.11	29.264
6	1½	19	2⅞	4242.92	29.466
6	1¾	19	3⅝	4271.83	29.665
6	2	19	4½	4300.85	29.867
6	2¼	19	5¼	4329.95	30.069
6	2½	19	6	4359.16	30.271
6	2¾	19	6¾	4388.47	30.475
6	3	19	7⅝	4417.87	30.619
6	3¼	19	8⅜	4447.37	30.884
6	3½	19	9⅛	4476.97	31.090
6	3¾	19	9⅞	4506.67	31.296
6	4	19	10¾	4536.47	31.503
6	4¼	19	11½	4566.36	31.710
6	4½	20	0¼	4596.35	31.919
6	4¾	20	1⅛	4626.44	32.114
6	5	20	1⅞	4656.63	32.337
6	5¼	20	2⅝	4686.92	32.548
6	5½	20	3⅛	4717.30	32.759
6	5¾	20	4¼	4747.79	32.970
6	6	20	5	4778.37	33.183
6	6¼	20	5¾	4809.05	33.396
6	6½	20	6½	4839.83	33.619
6	6¾	20	7⅜	4870.70	33.824
6	7	20	8⅛	4901.68	34.039
6	7¼	20	8⅞	4932.75	34.255
6	7½	20	9¾	4963.92	34.471
6	7¾	20	10½	4995.19	34.688
6	8	20	11¼	5026.26	34.906
6	8¼	21	0⅛	5058.02	35.125
6	8½	21	0⅞	5089.58	35.344
6	8¾	21	1⅝	5121.24	35.564
6	9	21	2⅜	5153.00	35.784
6	9¼	21	3¼	5184.86	36.006
6	9½	21	4	5216.82	36.227
6	9¾	21	4¾	5248.87	36.450
6	10	21	5½	5281.02	36.674
6	10¼	21	6⅜	5313.27	36.897
6	10½	21	7⅛	5345.62	37.122
6	10¾	21	7⅞	5378.07	37.347
6	11	21	8¾	5410.62	37.573
6	11¼	21	9½	5443.26	37.700
6	11½	21	10¼	5476.00	38.027
6	11¾	21	11	5508.84	38.256

Dia. in ft.	Dia. in in.	Circum. in ft.	Circum. in in.	Area in feet.
7	0	21	11⅞	38.4846
7	1	22	3	39.4060
7	2	22	6⅛	40.3388
7	3	22	9¼	41.2825
7	4	23	0⅜	42.2367
7	5	23	2⅛	43.2022
7	6	23	6¾	44.1787
7	7	23	11	45.1656
7	8	24	1⅛	46.1638
7	9	24	4⅛	47.1730
7	10	24	7¼	48.1926
7	11	24	10⅜	49.2236
8	0	25	1½	50.2656
8	1	25	4⅝	51.6178
8	2	25	7⅞	52.3816
8	3	25	11	53.4562
8	4	26	2⅛	54.5412
8	5	26	5¼	55.6377
8	6	26	8⅜	56.7451
8	7	26	11⅓	57.8628
8	8	27	2¾	58.9920
8	9	27	5¾	60.1321
8	10	27	9	61.2826
8	11	28	0⅛	62.4445
9	0	28	3¼	63.6174
9	1	28	6⅜	64.8006
9	2	28	9½	65.9951
9	3	29	0⅝	67.2007
9	4	29	3¾	68.4166
9	5	29	7	69.6440
9	6	29	10⅛	70.8823
9	7	30	1¼	72.1309
9	8	30	4⅜	73.3910
9	9	30	7½	74.6620
9	10	30	11⅝	75.9433
9	11	31	1¾	77.2362
10	0	31	5	78.5400
10	1	31	8⅛	79.8540
10	2	31	11¼	81.1795
10	3	32	2⅜	82.5190
10	4	32	5½	83.8627
10	5	32	8⅝	85.2211
10	6	32	11¾	86.5903
10	7	33	2⅞	87.9697
10	8	33	6⅛	80.3668
10	9	33	9¼	90.7627
10	10	34	0⅜	92.1749
10	11	34	3½	93.5986
11	0	34	6⅝	95.0334
11	1	34	9¾	96.4783
11	2	35	0⅞	97.9347
11	3	35	4⅛	99.4021
11	4	35	7¼	100.8797
11	5	35	10⅝	102.3689
11	6	36	1½	103.8601
11	7	36	4½	105.3794
11	8	36	7¾	106.9013
11	9	36	10⅞	108.4342
11	10	37	2¼	109.9772
11	11	37	5¼	111.5319
12	0	37	8⅜	113.0976
12	1	37	11½	114.6732
12	2	38	2⅝	116.2607
12	3	38	5¾	117.8590
12	4	38	8⅞	119.4674
12	5	39	0	121.0876
12	6	39	3¼	122.7187
12	7	39	6⅜	124.3593
12	8	39	9½	126.0127
12	9	40	0⅝	127.6765
12	10	40	3¾	129.3504
12	11	40	6⅞	131.0369
13	0	40	10	132.7326
13	1	41	1⅛	134.4391
13	2	41	4⅜	136.1574
13	3	41	7½	137.8867
13	4	41	10⅝	139.6260
13	5	42	1⅝	141.3771
13	6	42	4⅞	143.1391
13	7	42	8	144.9111
13	8	42	11⅛	146.6949
13	9	43	2¼	148.4896
13	10	43	5½	150.2943
13	11	43	8⅝	152.1109
14	0	43	11¾	153.9484
14	1	44	2⅞	155.7758
14	2	44	6	157.6250
14	3	44	9⅛	159.4852
14	4	45	0¼	161.3553
14	5	45	3½	163.2373
14	6	45	6⅝	165.1303
14	7	45	9¾	167.0331
14	8	46	0⅞	168.9479
14	9	46	4	170.8735
14	10	46	7⅛	172.8091
14	11	46	11¼	174.7565

Dia. in ft.	in.	Circum. in ft.	in.	Area in feet.	Dia. in ft.	in.	Circum. in ft.	in.	Area in feet.
15	0	47	1½	176.7150	17	0	53	4⅞	226.9806
15	1	47	4⅝	178.6832	17	1	53	8	229.2105
15	2	47	7¾	180.6624	17	2	53	11⅛	231.4625
15	3	47	10⅞	182.6545	17	3	54	2⅛	233.7055
15	4	48	2½	184.6555	17	4	54	5⅜	235.9682
15	5	48	5⅛	186.6684	17	5	54	8½	238.2430
15	6	48	8¼	188.6923	17	6	54	11⅝	240.5287
15	7	48	11⅜	190.7260	17	7	55	2⅞	242.8241
15	8	49	2⅝	192.7716	17	8	55	6	245.1316
15	9	49	5¾	194.8282	17	9	55	9⅛	247.4500
15	10	49	8⅞	196.8946	17	10	56	0¼	249.7781
15	11	50	0	198.9730	17	11	56	3½	252.1184
16	0	50	3⅛	201.0624	18	0	56	6½	254.4696
16	1	50	6¼	203.1615	18	1	56	9⅝	256.8303
16	2	50	9⅝	205.2726	18	2	57	0⅞	259.2033
16	3	51	0½	207.3946	18	3	57	4	261.5872
16	4	51	3¾	209.5264	18	4	57	7⅛	263.9807
16	5	51	6½	211.6703	18	5	57	10¼	266.3864
16	6	51	10	213.8251	18	6	58	1⅜	268.8031
16	7	52	1⅛	215.9896	18	7	58	4½	271.2293
16	8	52	4¼	218.1662	18	8	58	7⅝	273.6678
16	9	52	7⅜	220.3537	18	9	58	10¾	276.1171
16	10	52	10½	222.5510	18	10	59	2	278.5761
16	11	53	1⅝	224.7603	18	11	59	5½	281.0472

Diam.	Area.	Diam.	Area.	Diam.	Area.	Diam.	Area.	Diam.	Area.	Diam.	Area.	Diam.	Area.
1 in.	.7854	5 in.	19.635	9 in.	63.617	13 in.	132.732	17 in.	226.980	21 in.	346.361	25 in.	490.875
1/8	.9940	1/8	20.629	1/8	65.396	1/8	135.297	1/8	230.330	1/8	350.497	1/8	495.796
1/4	1.2271	1/4	21.647	1/4	67.200	1/4	137.886	1/4	233.705	1/4	354.657	1/4	500.741
3/8	1.4848	3/8	22.690	3/8	69.029	3/8	140.500	3/8	237.104	3/8	358.841	3/8	505.711
1/2	1.7671	1/2	23.758	1/2	70.882	1/2	143.139	1/2	240.528	1/2	363.051	1/2	510.706
5/8	2.0739	5/8	24.850	5/8	72.759	5/8	145.802	5/8	243.977	5/8	367.284	5/8	515.725
3/4	2.4052	3/4	25.967	3/4	74.662	3/4	148.489	3/4	247.450	3/4	371.543	3/4	520.769
7/8	2.7611	7/8	27.108	7/8	76.588	7/8	151.201	7/8	250.947	7/8	375.826	7/8	525.837
2 in.	3.1416	6 in.	28.274	10 in.	78.540	14 in.	153.938	18 in.	254.469	22 in.	380.133	26 in.	530.930
1/8	3.5465	1/8	29.464	1/8	80.515	1/8	156.699	1/8	258.016	1/8	384.465	1/8	536.047
1/4	3.9760	1/4	30.679	1/4	82.516	1/4	159.485	1/4	261.587	1/4	388.822	1/4	541.189
3/8	4.4302	3/8	31.919	3/8	84.540	3/8	162.295	3/8	265.182	3/8	393.203	3/8	546.356
1/2	4.9087	1/2	33.183	1/2	86.590	1/2	165.130	1/2	268.803	1/2	397.608	1/2	551.547
5/8	5.4119	5/8	34.471	5/8	88.664	5/8	167.989	5/8	272.447	5/8	402.038	5/8	556.762
3/4	5.9395	3/4	35.784	3/4	90.762	3/4	170.873	3/4	276.117	3/4	406.493	3/4	562.002
7/8	6.4918	7/8	37.122	7/8	92.865	7/8	173.782	7/8	279.811	7/8	410.972	7/8	567.267
3 in.	7.0686	7 in.	38.484	11 in.	95.033	15 in.	176.715	19 in.	283.529	23 in.	415.476	27 in.	572.556
1/8	7.6699	1/8	39.871	1/8	97.205	1/8	179.672	1/8	287.272	1/8	420.004	1/8	577.870
1/4	8.2957	1/4	41.282	1/4	99.402	1/4	182.654	1/4	291.039	1/4	424.557	1/4	583.208
3/8	8.9462	3/8	42.718	3/8	101.623	3/8	185.661	3/8	294.831	3/8	429.135	3/8	588.571
1/2	9.6211	1/2	44.178	1/2	103.869	1/2	188.692	1/2	298.648	1/2	433.731	1/2	593.958
5/8	10.320	5/8	45.663	5/8	106.139	5/8	191.748	5/8	302.489	5/8	438.363	5/8	599.370
3/4	11.044	3/4	47.173	3/4	108.434	3/4	194.828	3/4	306.355	3/4	443.014	3/4	604.807
7/8	11.793	7/8	48.707	7/8	110.753	7/8	197.933	7/8	310.245	7/8	447.699	7/8	610.268
4 in.	12.566	8 in.	50.265	12 in.	113.097	16 in.	201.062	20 in.	314.160	24 in.	452.390	28 in.	615.753
1/8	13.364	1/8	51.848	1/8	115.466	1/8	204.216	1/8	318.099	1/8	457.115	1/8	621.263
1/4	14.186	1/4	53.456	1/4	117.859	1/4	207.394	1/4	322.063	1/4	461.864	1/4	626.798
3/8	15.033	3/8	55.088	3/8	120.276	3/8	210.597	3/8	326.051	3/8	466.638	3/8	632.357
1/2	15.904	1/2	56.745	1/2	122.718	1/2	213.825	1/2	330.064	1/2	471.436	1/2	637.941
5/8	16.800	5/8	58.426	5/8	125.184	5/8	217.077	5/8	334.101	5/8	476.259	5/8	643.594
3/4	17.720	3/4	60.132	3/4	127.676	3/4	220.353	3/4	338.163	3/4	481.106	3/4	649.122
7/8	18.665	7/8	61.862	7/8	130.192	7/8	223.654	7/8	342.250	7/8	485.978	7/8	654.839
29 in.	660.521	29 1/8	666.227	29 1/4	671.958	29 3/8	677.714	29 1/2	683.494	29 3/4	695.128	30 in.	706.860

USE OF THE TABLE: To find the capacity of any cylindrical measure, from 1 inch diameter to 30 inches, take the inside diameter of the measure in inches, and multiply the area in the table which corresponds to the diameter by the depth in inches, and divide the products, if gills are required, by 7.2135; if pints, by 28.875; if quarts, by 57.75; and if gallons, by 231. If bushels are required (say in a tierce or barrel, after the mean diameter is obtained), multiply as above, and divide the product by 2150.42; the quotient is the number of bushels. Calling the diameters feet the areas are feet,—then, if a ship's water tank, steam boiler, etc., is 5⅛, or any number of feet and parts of feet in diameter, find the area in the table which corresponds in inches, multiply it by the length in feet, and multiply this result by the number of gallons in a cubic foot (7.4805), and the product is the answer in gallons. In any case where there are more figures in the divisor than in the dividend, add ciphers.

CAPACITY OF CANS ONE INCH DEEP.

USE OF THE TABLE.

Required the contents of a vessel, diameter 6 7-10 inches, depth 10 inches.

By the table a vessel 1 inch deep and 6 7-10 inches diameter contains .15 (hundredths) gallon, then 15 × 10 = 1.50, or 1 gallon and 2 quarts.

Required the contents of a can, diameter 19 8-10 inches, depth 30 inches.

By the table a vessel 1 inch deep and 19 8-10 inches diameter contains 1 gallon and .33 (hundredths), then 1.33 × 30 = 39.90, or nearly 40 gallons.

Required the depth of a can whose diameter is 12 2-10 inches, to contain 16 gallons.

By the table a vessel 1 inch deep and 12 2-10 inches diameter contains .50 (hundredths) gallon, then 16 ÷ .50 = 32 inches, the depth required.

Diameter.		$^1/_{10}$	$^2/_{10}$	$^3/_{10}$	$^4/_{10}$	$^5/_{10}$	$^6/_{10}$	$^7/_{10}$	$^8/_{10}$	$^9/_{10}$
3	.03	.03	.03	.03	.03	.04	.04	.04	.04	.05
4	.05	.05	.05	.05	.06	.06	.07	.07	.07	.08
5	.08	.08	.08	.08	.09	.10	.10	.11	.11	.11
6	.12	.12	.12	.13	.13	.14	.14	.15	.15	.16
7	.16	.17	.17	.18	.18	.19	.19	.20	.20	.21
8	.21	.22	.22	.23	.23	.24	.25	.25	.26	.26
9	.27	.28	.28	.29	.30	.30	.31	.31	.32	.33
10	.34	.34	.35	.36	.36	.37	.38	.38	.39	.40
11	.41	.41	.42	.43	.44	.44	.45	.46	.47	.48
12	.48	.49	.50	.51	.52	.53	.53	.54	.55	.56
13	.57	.58	.59	.60	.60	.61	.62	.63	.64	.65
14	.66	.67	.68	.69	.70	.71	.72	.73	.74	.75
15	.76	.77	.78	.79	.80	.81	.82	.83	.84	.85
16	.87	.88	.89	.90	.91	.92	.93	.94	.95	.97
17	.98	.99	1.005	1.017	1.028	1.040	1.051	1.063	1.075	1.086
18	1.101	1.113	1.125	1.138	1.150	1.162	1.170	1.187	1.200	1.211
19	1.227	1.240	1.253	1.266	1.279	1.292	1.304	1.317	1.330	1.343
20	1.360	1.373	1.385	1.400	1.414	1.428	1.441	1.455	1.478	1.482
21	1.499	1.513	1.527	1.542	1.556	1.570	1.585	1.600	1.612	1.630
22	1.645	1.660	1.675	1.696	1.705	1.720	1.735	1.750	1.770	1.780
23	1.798	1.814	1.830	1.845	1.861	1.876	1.892	1.908	1.923	1.940
24	1.958	1.974	1.991	2.007	2.023	2.040	2.056	2.072	2.096	2.105
25	2.125	2.142	2.159	2.176	2.193	2.120	2.227	2.244	2.261	2.280
26	2.298	2.316	2.333	2.351	2.369	2.386	2.404	2.422	2.440	2.460
27	2.478	2.496	2.515	2.533	2.552	2.570	2.588	2.607	2.625	2.643
28	2.665	2.684	2.703	2.722	2.741	2.764	2.780	2.800	2.820	2.836
29	2.859	2.879	2.898	2.918	2.938	2.958	2.977	2.997	3.017	3.036
30	3.060	3.080	3.100	3.121	3.141	3.162	3.182	3.202	3.223	3.245
31	3.267	3.288	3.309	3.330	3.351	3.372	3.393	3.414	3.436	3.457
32	3.481	3.503	3.524	3.543	3.568	3.590	3.612	3.633	3.655	3.589
33	3.702	3.725	3.747	3.773	3.795	3.814	3.837	3.860	3.882	3.904
34	3.930	3.953	3.976	4.003	4.022	4.046	4.070	4.092	4.115	4.140
35	4.165	4.188	4.212	4.236	4.260	4.284	4.307	4.331	4.355	4.380
36	4.406	4.430	4.455	4.483	4.503	4.528	4.553	4.577	4.602	4.626
37	4.654	4.679	4.704	4.730	4.755	4.780	4.805	4.834	4.855	4.880
38	4.909	4.935	4.961	4.987	5.012	5.038	6.064	5.090	5.120	5.142
39	5.171	5.197	5.224	5.250	5.277	5.304	5.330	5.357	5.383	5.410
40	5.440	5.467	5.491	5.521	5.548	5.576	5.603	5.630	5.657	5.684

RULES FOR CALCULATING CIRCUM-
FERENCES.

1st. Multiply the given diameter by 22, and divide the product by 7; or 2d, divide 22 by 7 and multiply the diameter by the quotient; or 3d, multiply the diameter by 3.1416; or 4th, multiply the diameter by 3 and add 1 inch for every 7 of the diameter, or about ⅛ inch for every 1. For example: If the given diameter be 15 inches, by the first rule the circumference would be 47 1-7 inches; by the second, 47 1-7 inches; by the third, 47.1240 inches; by the fourth, 47⅛ inches; by the table, 47⅛ inches. It will be seen that the result is not just the same by the several rules, yet either is near enough for general use and practice.

WEIGHT OF WATER.

1	cubic inch..............	is equal to	.03617	pound.
12	cubic inches............	is equal to	.434	pound.
1	cubic foot.............	is equal to	62.5	pounds.
1	cubic foot.............	is equal to	7.50	U. S. gallons.
1.8	cubic feet.............	is equal to	112.00	pounds.
35.84	cubic feet.............	is equal to	2240.00	pounds.
1	cylindrical inch.........	is equal to	.02842	pound.
12	cylindrical inches.......	is equal to	.341	pound.
1	cylindrical foot.........	is equal to	49.10	pounds.
1	cylindrical foot.........	is equal to	6.00	U. S. gallons.
2.282	cylindrical feet.........	is equal to	112.00	pounds.
45.64	cylindrical feet.........	is equal to	2240.00	pounds.
13.43	United States gallons...	is equal to	112.00	pounds.
268.8	United States gallons...	is equal to	2240.00	pounds.

Center of pressure is at two-thirds depth from surface.

TO FIND NUMBER OF BARRELS IN
CISTERNS.

The following table shows the number of barrels (31½ gallons) contained in cisterns of various diameters, from 5 to 30 feet, and of depths ranging from 5 to 20 feet.

To use the table, find the required depth in the side column, and then follow along the line to the column which has the required diameter at the top. Thus, with a cistern 6 feet deep and 16 feet in diameter, we find 6 in the second line, and then follow along until column 16 is reached, when we find that the contents is 286.5 barrels.

NUMBER OF BARRELS (31½ GALLONS) IN CISTERNS AND TANKS.

Diameter in feet.

Depth in feet.	5	6	7	8	9	10	11	12	13
5	23.3	33.6	45.7	59.7	75.5	93.2	112.8	134.3	157.6
6	28.0	40.3	54.8	71.7	90.6	111.9	135.4	161.1	189.1
7	32.7	47.0	64.0	83.6	105.7	130.6	158.0	188.0	220.6
8	37.3	53.7	73.1	95.5	120.9	149.2	180.5	214.8	252.1
9	42.0	60.4	82.2	107.4	136.0	167.9	203.1	241.7	283.7
10	46.7	67.1	91.4	119.4	151.1	186.5	225.7	268.6	315.2
11	51.3	73.9	100.5	131.3	166.2	205.1	248.2	295.4	346.7
12	56.0	80.6	109.7	143.2	181.3	223.8	270.8	322.3	378.2
13	60.7	87.3	118.8	155.2	196.4	242.4	293.4	349.1	409.7
14	65.3	94.0	127.9	167.1	211.5	261.1	315.9	376.0	441.3
15	70.0	100.7	137.1	179.0	226.6	289.8	338.5	402.8	472.8
16	74.7	107.4	146.2	191.0	241.7	298.4	361.1	429.7	504.3
17	79.3	114.1	155.4	202.9	256.8	317.0	383.6	456.6	535.8
18	84.0	120.9	164.5	214.8	272.0	335.7	406.2	483.4	567.3
19	88.7	127.6	173.6	226.8	287.0	354.3	428.8	510.3	598.0
20	93.3	134.3	182.8	238.7	302.1	373.0	451.3	537.1	630.4

Diameter in feet.

Depth in feet.	14	15	16	17	18	19	20	21	22
5	182.8	209.8	238.7	269.5	302.1	336.6	373.0	411.2	451.3
6	219.3	251.8	286.5	323.4	362.6	404.0	447.6	493.5	541.6
7	255.9	293.7	334.2	377.3	423.0	471.3	522.2	575.7	631.9
8	292.4	335.7	382.0	431.2	483.4	538.6	596.8	658.0	722.1
9	329.0	377.7	429.7	485.1	543.8	605.9	671.4	740.2	812.4
10	365.5	419.6	477.4	539.0	604.3	673.3	746.0	822.5	902.7
11	402.1	461.6	525.2	592.9	667.7	740.6	820.6	904.7	992.9
12	438.6	503.5	572.9	646.8	725.1	807.9	895.2	987.0	1083.2
13	475.2	545.5	620.7	700.7	785.5	875.2	969.8	1069.2	1173.5
14	511.8	587.5	668.2	754.6	846.6	942.6	1044.4	1151.5	1263.7
15	548.3	629.4	716.2	308.5	906.0	1009.9	1119.0	1233.7	1354.0
16	584.9	671.4	773.9	862.4	966.8	1077.2	1193.6	1315.9	1444.3
17	621.4	713.4	811.6	916.3	1027.2	1144.6	1268.2	1398.2	1534.5
18	658.0	755.3	859.4	970.2	1087.7	1211.9	1342.8	1480.4	1624.8
19	694.5	797.3	907.1	1024.1	1148.1	1279.2	1417.4	1562.7	1715.1
20	731.1	839.3	954.9	1078.0	1208.5	1346.5	1492.0	1644.9	1805.3

Diameter in feet.

Depth in feet.	23	24	25	26	27	28	29	30
5	493.3	537.1	582.8	630.4	679.8	731.1	784.2	839.3
6	592.0	644.5	699.4	756.5	815.8	877.3	941.1	1007.1
7	690.6	752.0	815.9	882.5	951.7	1023.5	1097.9	1175.0
8	789.3	859.4	932.5	1008.6	1087.7	1169.7	1254.8	1342.8
9	887.9	966.8	1049.1	1134.7	1223.6	1316.0	1411.6	1510.7
10	986.6	1074.2	1165.6	1260.8	1359.6	1462.2	1568.2	1678.5
11	1085.2	1181.7	1282.2	1386.8	1495.6	1608.7	1723.0	1846.4
12	1183.9	1289.1	1398.7	1512.9	1631.5	1754.6	1882.2	2014.2
13	1282.6	1396.5	1515.3	1639.0	1767.5	1900.8	2039.0	2182.0
14	1381.2	1503.9	1631.9	1765.1	1903.4	2047.1	2195.9	2343.9
15	1479.9	1611.4	1748.4	1891.1	2039.4	2193.3	2352.7	2517.8
16	1578.5	1718.8	1865.0	2017.2	2175.4	2339.5	2509.6	2685.6
17	1677.2	1826.2	1981.6	2143.3	2311.3	2485.7	2666.4	2853.5
18	1775.9	1933.6	2098.1	2269.4	2447.3	2631.9	2823.3	3021.3
19	1874.5	2041.1	2214.7	2395.4	2583.2	2778.1	2980.1	3189.2
20	1973.2	2148.5	2321.2	2521.5	2719.2	2924.4	3137.0	3357.0

For tanks that are tapering the diameter may be measured four-tenths from large end.

TABLE SHOWING THE PRESSURE OF WATER PER SQUARE INCH, DUE TO DIFFERENT HEADS, FROM 1 TO 250 FEET.

Head.	Pressure in lbs.	Head.	Pressure in lbs.	Head.	Pressure in lbs.
1	.4335	19	8.237	37	16.04
2	.8670	20	8.670	38	16.47
3	1.300	21	9.104	39	16.91
4	1.734	22	9.537	40	17.34
5	2.167	23	9.971	50	21.67
6	2.601	24	10.40	100	43.35
7	3.035	25	10.84	110	47.68
8	3.408	26	11.27	120	52.02
9	3.902	27	11.70	130	56.36
10	4.335	28	12.14	140	60.69
11	4.768	29	12.57	150	65.03
12	5.202	30	13.00	160	69.36
13	5.636	31	13.44	170	73.70
14	6.069	32	13.87	180	78.03
15	6.503	33	14.31	190	82.36
16	6.936	34	14.74	200	86.70
17	7.370	35	15.17	225	97.41
18	7.803	36	15.60	250	108.37

MEASURES OF CAPACITY AND WEIGHT.

MEASURES OF WEIGHT.—AVOIRDUPOIS.—16 drams equal 1 ounce; 16 ounces 1 pound; 112 pounds 1 hundredweight; 20 hundredweights 1 ton. TROY.—24 grains 1 pennyweight; 20 pennyweights 1 ounce; 12 ounces 1 pound. APOTHECARIES'.—20 grains equal 1 scruple; 3 scruples 1 dram; 8 drams 1 ounce; 12 ounces 1 pound.

MEASURES OF CAPACITY (DRY).—2150.42 cubic inches equal 1 United States (or Winchester) bushel; the dimensions of which are 18½ inches diameter inside, 19½ inches outside and 8 inches deep; 2747.70 cubic inches equal 1 heaped bushel, the cone of which must not be less than 6 inches high.

MEASURES OF CAPACITY (LIQUIDS).—231 cubic inches equal 1 United States standard gallon; 277.274 cubic inches equal 1 Imperial (British) gallon; 31½ United States gallons equal 1 barrel; 42 gallons equal 1 tierce; 63 gallons equal 1 hogshead; 84 gallons equal 1 puncheon; 126 gallons equal 1 pipe; 252 gallons equal 1 tun.

FRENCH MEASURES OF FREQUENT REFERENCE, COMPARED WITH U. S. MEASURES.—Meter, 3.28 feet; Decimeter (1-10 meter), 3.94 inches; Centimeter, .4 inch; Millimeter, .04 inch; Hectoliter, 26.42 gallons; Liter, 2.11 pints; Kilogram, 2.2 pounds.

WEIGHTS OF VARIOUS SUBSTANCES.—POUNDS AVOIRDUPOIS.—1 cubic foot of bricks weighs 124 pounds; 1 do. of sand or loose earth, 95; 1 do. of cork, 15; 1 do. of granite, 170; 1 do. of cast iron, 450; 1 do. of wrought iron, 485; 1 do. of steel, 490; 1 do. of copper, 555; 1 do. lead, 709; 1 do. brass, 520; 1 do. tin, 459; 1 do. white pine, 30; 1 do. oak, 48; 1 do. sea water, 64.08; 1 do. fresh, 62.35; 1 do. air, 0765.

SIZES OF TIN WARE IN THE FORM OF FRUSTUM OF A CONE.

PANS.

Size.	Diam. of top.	Diam. of bot.	Hight.	Size.	Diam. of top.	Diam. of bot.	Hight.
20 qt.	19½ in.	13 in.	8 in.	2 qt.	9 in.	6 in.	3¾ in.
16 "	18 "	11¼ "	6¼ "	3 pt.	8¼ "	5¾ "	2¾ "
14 "	15¼ "	9¼ "	6¼ "	1 "	6¼ "	4 "	2¾ "
10 "	14¾ "	11 "	4⅛ "	Pie	9 "	7½ "	1¾ "
6 "	12¾ "	9 "	4 "				

DISH KETTLES AND PAILS.

Size.	Diam. of top.	Diam. of bot.	Hight.	Size.	Diam. of top.	Diam. of bot.	Hight.
14 qt.	13 in.	9 in.	9 in.	6 qt.	9¼ in.	5½ in.	6½ in.
10 "	11½ "	7 "	8 "	2 "	6¼ "	4 "	4 "

COFFEE POTS.

Size.	Diam. of top.	Diam. of bot.	Hight.	Size.	Diam. of top.	Diam. of bot.	Hight.
1 gal.	4 in.	7 in.	8½ in.	3 qt.	3½ in.	6 in.	8½ in.

WASH BOWLS.

Size	Diam. of top.	Diam. of bot.	Hight.
Large wash bowl	11 in.	5¾ in.	5 in.
Cullender	11 "	5¾ "	5 "
Small wash bowl	9½ "	5½ "	3¾ "
Milk strainer	9½ "	5½ "	3¾ "

DIPPERS.

Size.	Diam. of top.	Diam. of bot.	Hight.	Size.	Diam. of top.	Diam. of bot.	Hight.
½ gal.	6½ in.	4 in.	4 in.	1 pt.	4¼ in.	3¾ in.	2¾ in.

MEASURES.

Size.	Diam. of top.	Diam. of bot.	Hight.	Size.	Diam. of top.	Diam. of bot.	Hight.
1 gal.	5½ in.	6⅛ in.	9¼ in.	1 pt.	2⅛ in.	3¾ in.	4¼ in.
½ "	4 "	4⅞ "	8 "	½ "	2⅜ "	2⅞ "	3⅛ "
1 qt.	3½ "	4 "	5¾ "				

DRUGGISTS' AND LIQUOR DEALERS' MEASURES.

Size.	Diam. of top.	Diam. of bot.	Hight.	Size.	Diam. of top.	Diam. of bot.	Hight.
5 gal.	8 in.	13½ in.	12⅜ in.	½ gal.	3¼ in.	6⅝ in.	6 in.
3 "	7 "	11½ "	10⅛ "	1 qt.	2½ "	5⅛ "	4⅞ "
2 "	6 "	10½ "	8⅜ "	1 pt.	2 "	4 "	4 "
1 "	3¾ "	8¾ "	7½ "	½ "	1¾ "	3¾ "	3⅛ "

TABLE OF EFFECTS UPON BODIES BY HEAT.

	Degrees F.
Cast iron thoroughly melts at	2,228
Gold melts at	1,913
Silver melts at	1,733
Copper melts at	1,929
Brass melts at	1,873
Zinc melts at	779
Lead melts at	618
Bismuth melts at	506
Tin melts at	444
Tin and lead, equal parts, melt at	418
Tin 2 parts, bismuth 5 and lead 3, melt at	199

PRACTICAL RECEIPTS.

SOLDERS.

SOLDER FOR GOLD.

Gold, 6 pennyweights; silver, 1 pennyweight; copper, 2 pennyweights.

SOLDER FOR SILVER, FOR THE USE OF JEWELERS.

Fine silver, 19 pennyweights; copper, 1 pennyweight; sheet brass, 10 pennyweights.

WHITE SOLDER FOR SILVER.

Silver, 1 ounce; tin, 1 ounce.

WHITE SOLDER FOR RAISED BRITANNIA WARE.

Tin, 100 pounds; copper, 3 ounces; to make it free, add lead, 3 ounces.

BEST SOFT SOLDER FOR CAST BRITANNIA WARE.

Tin, 8 pounds; lead, 5 pounds.

YELLOW SOLDER FOR BRASS OR COPPER.

Copper, 1 pound; zinc, 1 pound.

YELLOW SOLDER FOR BRASS OR COPPER.

(Stronger than the last.)   Copper, 32 pounds; zinc, 29 pounds; tin, 1 pound.

SOLDER FOR COPPER.

Copper, 10 pounds; zinc, 9 pounds.

BLACK SOLDER.

Copper, 2 pounds; zinc, 3 pounds; tin, 2 ounces.

BLACK SOLDER.

Sheet brass, 20 pounds; tin, 6 pounds; zinc, 1 pound.

SILVER SOLDER FOR PLATED METAL.

Fine silver, 1 ounce; brass, 10 pennyweights.

PLUMBERS' SOLDER.

Lead, 2; tin, 1 part.

TINMEN'S SOLDER.

Lead, 1; tin, 1 part.

PEWTERERS' SOLDER.

Tin, 2; lead, 1 part.

HARD SOLDER.

Copper, 2; zinc, 1 part.

SOLDER FOR STEEL JOINTS.

Silver, 19 pennyweights; copper, 1 pennyweight; brass, 2 pennyweights.   Melt under a coat of charcoal dust.

SOFT GOLD SOLDER

Is composed of 4 parts gold, 1 of silver and 1 of copper. It can be made softer by adding brass, but the solder becomes more liable to oxidize.

CEMENT FOR MENDING EARTHEN AND GLASS WARE.

1. Heat the article to be mended a little above boiling water heat, then apply a thin coating of gum shellac on both surfaces of the broken vessel, and when cold it will be as strong as it was originally. 2. Dissolve gum shellac in alcohol, apply the solution and bind the parts firmly together until the cement is perfectly dry.

CEMENT FOR STONE WARE.

Another cement in which an analogous substance, the curd of milk, is employed, is made by boiling slices of skim milk cheese into a gluey consistence in a great quantity of water, and then incorporating it with quicklime on a slab with a muller, or in a marble mortar. When this compound is applied warm to broken edges of stone ware, it unites them very firmly after it is cold.

IRON RUST CEMENT

Is made from 50 to 100 parts of iron borings, pounded and sifted, mixed with 1 part of sal ammoniac, and when it is to be applied, moistened with as much water as will give it a pasty consistency. Another composition of the same kind is made by mixing 4 parts of fine borings or filings of iron, 2 parts of potters' clay and 1 part of pounded potsherds, and making them into a paste with salt and water.

CEMENT FOR IRON TUBES, BOILERS, ETC.

Finely powdered iron, 66 parts; sal ammoniac, 1 part; water, a sufficient quantity to form a paste.

CEMENT FOR IVORY, MOTHER OF PEARL, ETC.

Dissolve 1 part of isinglass and 2 of white glue in 30 of water, strain and evaporate to 6 parts. Add 1-30 part of gum mastic, dissolve in ½ part of alcohol and 1 part of white zinc. When required for use warm and shake up.

CEMENT FOR HOLES IN CASTINGS.

The best cement for this purpose is made by mixing 1 part of sulphur in powder, 2 parts of sal ammoniac and 80 parts of clean powdered iron turnings. Sufficient water must be added to make it into a thick paste, which should be pressed into the holes or seams which are to be filled up. The ingredients composing this cement should be kept separate and not mixed until required for use. It is to be applied cold, and the casting should not be used for two or three days afterward.

CEMENT FOR COPPERSMITHS AND ENGINEERS.

Boiled linseed oil and red lead mixed together into a putty is often used by coppersmiths and engineers to secure joints. The washers of leather or cloth are smeared with this mixture in a pasty state.

A CHEAP CEMENT.

Melted brimstone, either alone or mixed with rosin and brick dust, forms a tolerably good and very cheap cement.

PLUMBERS' CEMENT

Consists of black rosin, 1 part; brick dust, 2 parts; well incorporated by a melting heat.

CEMENT FOR BOTTLE CORKS.

The bituminous or black cement for bottle corks consists of pitch hardened by the addition of rosin and brick dust.

CHINA CEMENT.

Take the curd of milk, dried and powdered, 10 ounces; quicklime, 1 ounce; camphor, 2 drams. Mix and keep in closely stopped bottles. When used, a portion is to be mixed with a little water into a paste, to be applied quickly

CEMENT FOR LEATHER.

A mixture of India rubber and shellac varnish makes a very adhesive leather cement. A strong solution of common isinglass, with a little diluted alcohol added to it, makes an excellent cement for leather.

MARBLE CEMENT.

Take plaster of paris and soak it in a saturated solution of alum, then bake the two in an oven, the same as gypsum is baked to make it plaster of paris; after which they are ground to powder. It is then used as wanted, being mixed up with water like plaster and applied. It sets into a very hard composition capable of taking a very high polish. It may be mixed with various coloring minerals to produce a cement of any color capable of imitating marble.

CEMENT FOR MARBLE WORKERS AND COPPERSMITHS.

White of an egg alone, or mixed with finely sifted quicklime, will answer for uniting objects which are not exposed to moisture. The latter combination is very strong and is much employed for joining pieces of spar and marble ornaments. A similar composition is used by coppersmiths to secure the edges and rivets of boilers, only bullock's blood is the albuminous matter used instead of white of egg.

TRANSPARENT CEMENT FOR GLASS.

Dissolve 1 part of india rubber in 64 of chloroform, then add gum mastic in powder 14 to 24 parts, and digest for two days with frequent shaking. Apply with camel's hair brush.

CEMENT TO MEND IRON POTS AND PANS.

Take 2 parts of sulphur, and 1 part, by weight, of fine black lead; put the sulphur in an old iron pan, holding it over the fire until it begins to melt, then add the lead, stir well until all is mixed and melted, then pour out on an iron plate or smooth stone. When cool, break into small pieces. A sufficient quantity of this compound being placed upon the crack of the iron pot to be mended, can be soldered by a hot iron in the same way a tinsmith solders his sheets. If there is a small hole in the pot, drive a copper rivet in it and then solder over it with this cement.

CEMENT TO RENDER CISTERNS AND CASKS WATER TIGHT.

An excellent cement for resisting moisture is made by incorporating thoroughly 8 parts of melted glue, of the consistence used by carpenters, with 4 parts of linseed oil, boiled into varnish with litharge. This cement hardens in about 48 hours and renders the joints of wooden cisterns and casks air and water tight. A compound of glue with one-quarter its weight of Venice turpentine, made as above, serves to cement glass, metal and wood to one another. Fresh made cheese curd and old skim milk cheese, boiled in water to a slimy consistency, dissolved in a solution of bicarbonate of potash are said to form a good cement for glass and porcelain. The gluten of wheat, well prepared, is also a good cement. White of eggs with flour and water, well mixed, and smeared over linen cloth, forms a ready lute for steam joints in small apparatus.

A GOOD CEMENT.

Shellac, dissolved in alcohol or in a solution of borax, forms a pretty good cement.

CEMENT FOR REPAIRING FRACTURED BODIES OF ALL KINDS.

White lead ground upon a slab with linseed oil varnish and kept out of contact of air affords a cement capable of repairing fractured bodies of all kinds. It requires a few weeks to harden. When stone and iron are to be cemented together, a compound of equal parts of sulphur with pitch answers very well.

CEMENT FOR CRACKS IN WOOD.

Make a paste of slaked lime 1 part, rye meal 2 parts, with a sufficient quantity of linseed oil. Or dissolve 1 part of glue in 16 parts of water, when almost cool stir in sawdust and prepared chalk a sufficient quantity. Or oil varnish thickened with a mixture of equal parts of white lead, red lead, litharge and chalk.

CEMENT FOR JOINING METALS AND WOOD.

Melt rosin and stir in calcined plaster until reduced to a paste, to which add boiled oil a sufficient quantity to bring it to the consistence of honey; apply warm. Or, melt rosin 180 parts and stir in burnt umber 30, calcined plaster 15 and boiled oil 8 parts.

GAS FITTERS' CEMENT.

Mix together resin 4½ parts, wax 1 part, and Venetian red 3 parts.

IMPERVIOUS CEMENT FOR APPARATUS, CORKS, ETC.

Zinc white rubbed up with copal varnish to fill up the indentures; when dry, to be covered with the same mass somewhat thinner, and lastly with copal varnish alone.

CEMENT FOR FASTENING BRASS TO GLASS VESSELS.

Melt rosin 150 parts, wax 30, and add burnt ocher 30 and calcined plaster 2 parts. Apply warm.

CEMENT FOR FASTENING BLADES, FILES, ETC.

Shellac 2 parts, prepared chalk 1, powdered and mixed. The opening for the blade is filled with this powder, the lower end of the iron heated and pressed in.

HYDRAULIC CEMENT PAINT.

If hydraulic cement be mixed with oil, it forms a first rate anti-combustible and excellent water proof paint for roofs of buildings, outhouses, walls, &c.

TO STOP A LEAKY ROOF.

Twenty-five pounds yellow ocher, 1 pound litharge, 6 pounds black lead, 1 pound fine salt; boil well in oil. Soak strips of cloth in the above and paste over the seams. Good where solder is not practicable.

FLUX FOR SOLDERING TIN ROOF.

One part rosin and 2 parts binnacle oil mixed hot and used the same as rosin alone; or, cut with alcohol 1 pint as much rosin as possible and put on with a swab. Either good when the wind blows. Or saponified or red oil used with a swab along the seams. Solder flows more freely than with rosin alone.

SOLDERING FLUID OR FLUX.

Prussiate of potash, borax and copperas, each 1 dram; sal ammoniac ½ ounce, muriatic acid 3½ ounces, well mixed, then add as much zinc as it will dissolve. Add 1 pint or more water according to strength required.

ANOTHER.

Sal ammoniac and borax, each 1 dram; chloride of zinc 1 ounce, water 1 pint. It will not eat copper or tarnish tin. Use less water and it will be stronger.